TABLE OF CONTENTS

This handbook describes solar heating applications for livestock buildings in the North Central Region. Information on collecting and using solar energy as heat energy and economic and design considerations are included. Solar radiation data by city and latitude and a glossary are at the end of this book.

SOLAR FUNDAMENTALS

Solar energy comes from thermonuclear reactions in the core of the sun. The energy is mostly shortwave radiation emitted into space in all directions. When it strikes a material, the radiation can be reflected, transmitted, or absorbed. The absorbed fraction causes the material to heat.

Solar energy is nonpolluting—it has no undesirable byproducts. The energy itself is free, but the equipment required to collect and use it is not. The relative costs of solar collection equipment compared with fossil fuel (coal, gas, and oil) costs have slowed solar development. But solar energy collection is already economically feasible for a few applications, and if fossil fuel costs continue to increase, more solar applications may become feasible. Solar energy can reduce our dependence on fossil fuels, although its availability is too variable and too limited to completely replace fossil fuels.

This discussion of solar energy fundamentals gives basic information on solar energy and on collecting and using it as heat energy. Agricultural applications, especially those related to livestock buildings, are emphasized. A study of solar energy fundamentals will help you:

- Determine how much solar energy is available in your area.
- Determine if solar energy is realistic for your operation.
- Select the best type of commercial or home-built collector for your situation.
- Calculate the size of solar collector you need.
- Determine the size and type of solar energy storage system best suited to your needs.
- Plan or design a collector to meet your needs.

Available Solar Energy

The quantity of solar energy, or solar radiation, reaching a surface decreases with greater distance from the sun. The sun is not at the center of the earth's orbit, so the earth's distance from the sun varies during the year, Fig 1. The average value for the solar radiation reaching the outside of the earth's atmosphere—called the solar constant—is about 428 Btu/hr-ft^2. Total average daily radiation received outside the earth's atmosphere is 24 hr x 428 Btu/hr-ft^2 = 10,272 Btu/day-ft^2. Only a fraction of this energy is available to solar collectors on the earth's surface. The actual amount of solar energy available to collectors on the earth depends on time of day, time of year, latitude, collector tilt angle, and weather.

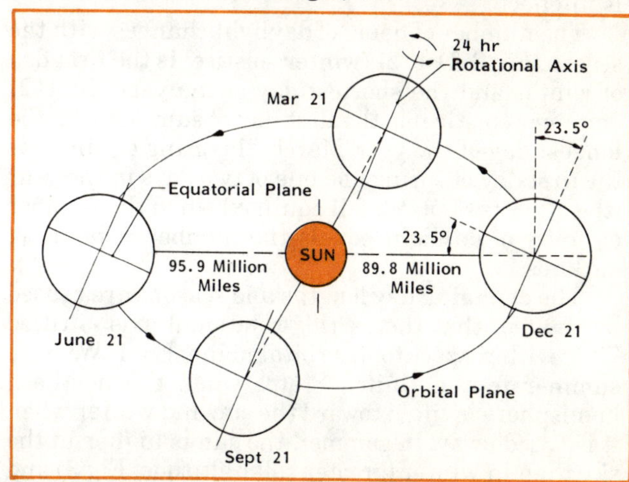

Fig 1. Earth's motion around the sun.
Solar year is one revolution of 365¼ days.

Day-Night Cycle and Season

Because of the earth's rotation, solar collectors on the earth's surface do not receive energy 24 hr/day. On sunny days, solar radiation increases from zero just before dawn to a maximum at solar noon and decreases to zero again at dusk, Fig 2.

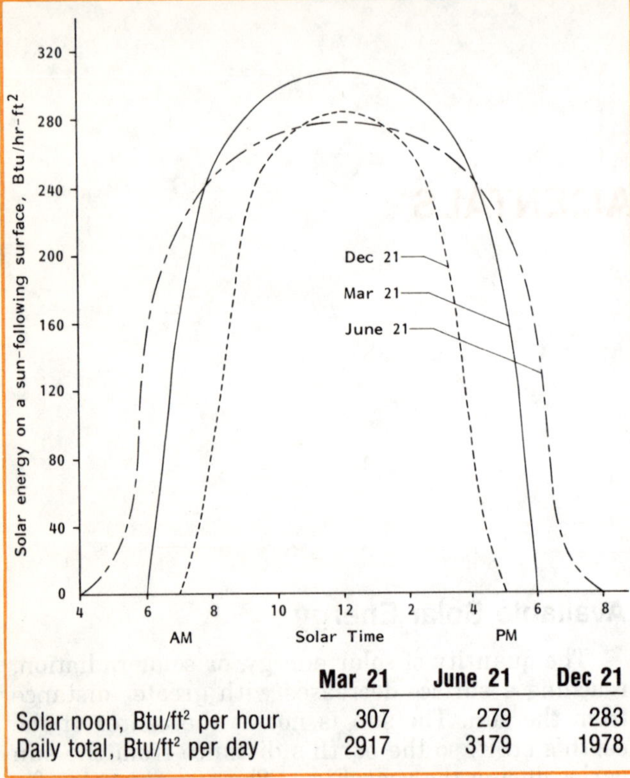

	Mar 21	June 21	Dec 21
Solar noon, Btu/ft² per hour	307	279	283
Daily total, Btu/ft² per day	2917	3179	1978

Fig 2. Hourly clear day radiation.
At 40° north latitude on a surface that follows the sun across the sky.

Note that solar time (used in Fig 2 and Table 26, appendix) is not the same as local clock time. At solar noon, the sun reaches its highest point in the sky for the day. Solar noon varies from local clock noon depending on time of year, where you live in the time zone (longitude), and whether Daylight Savings Time is in effect.

The number of hours of daylight changes with the season, Fig 2. Dec. 21 (winter solstice) is the first day of winter and the shortest day of the year. June 21 (summer solstice) is the first day of summer and the longest day of the year. March 21 (spring equinox) is the first day of spring and one of two days in the year (the other is Sept. 21, fall equinox) when the number of hours of daylight equals the number of hours of darkness.

The changing day lengths and seasons are caused by the fact that the earth's rotational axis is tilted 23.5° with respect to its orbital plane, Fig 1. We have summer in the United States when the northern hemisphere is tilted toward the sun and winter when it is tilted away. In summer, the sun is higher in the sky than in winter (greater solar altitude, Fig 3) and it rises and sets further to the north (greater solar azimuth, Fig 3). Because the sun is higher in the sky

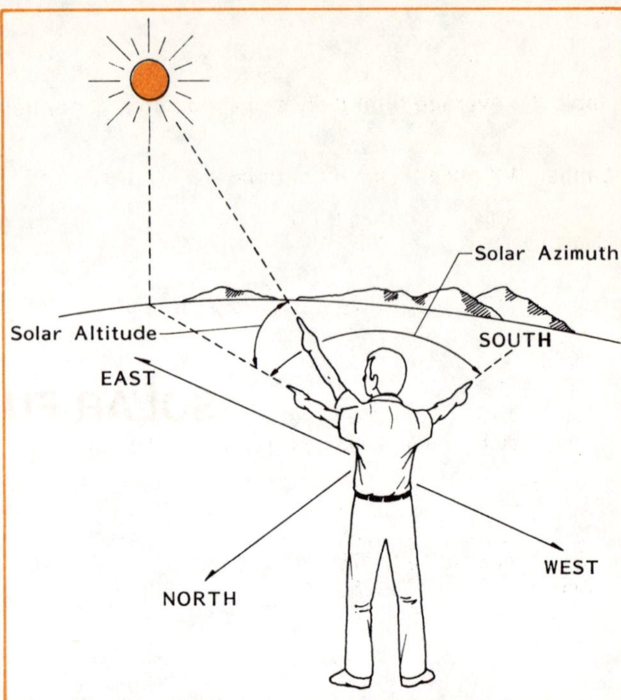

Fig 3. Solar altitude and solar azimuth angles.

(see Atmospheric Effects) and there are more hours of daylight, more daily total solar energy is received on surfaces exposed to the sun in summer. See the sun-following surface curve in Fig 4.

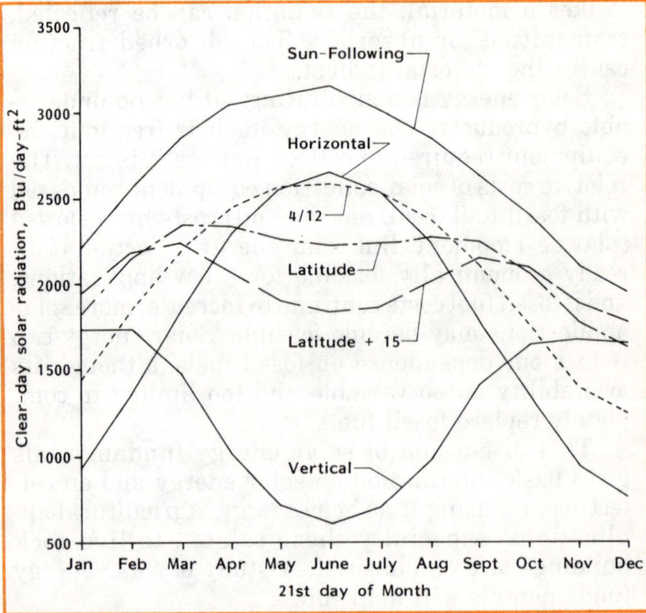

Fig 4. Clear day solar radiation.
South-facing surface. 40° north latitude. Curve labels are collector tilt angles measured from horizontal.

Surface Orientation

A sun-following surface—one that tracks or follows the sun's movement across the sky—receives more solar energy than surfaces with any other orientation, Fig 4. The sun-following surface is always perpendicular to the sun's rays, so the angle of inci-

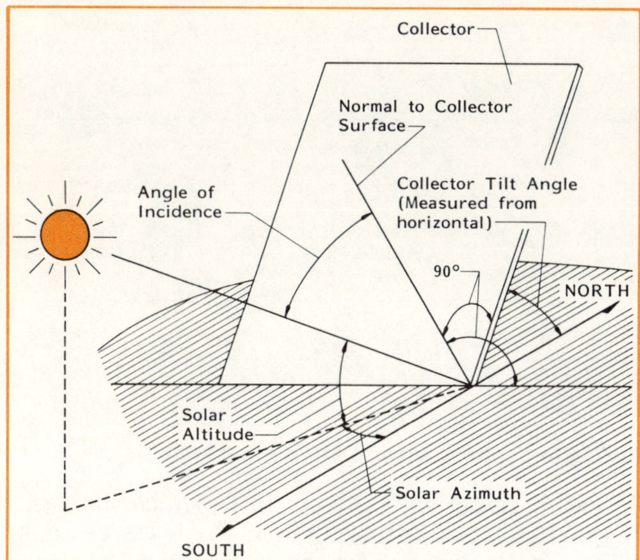

Fig 5. Solar angle of incidence on south-facing surface.
The angle of incidence is measured from the sun to normal to surface.

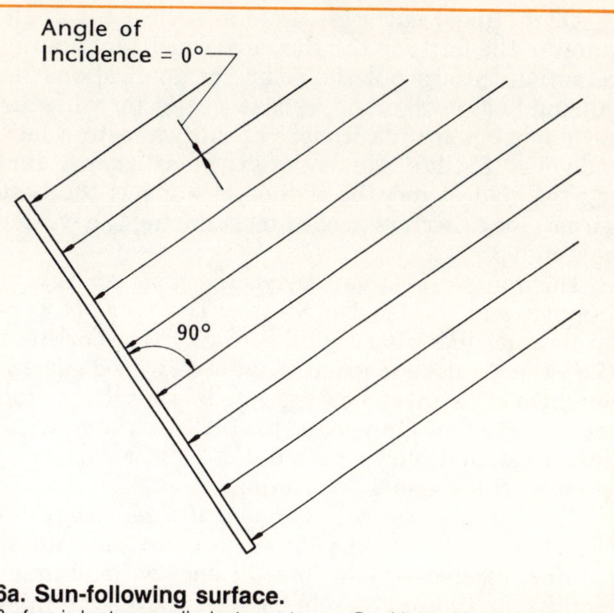

6a. Sun-following surface.
Surface is kept perpendicular to sun's rays. Provides maximum interception and minimum reflection of radiation.

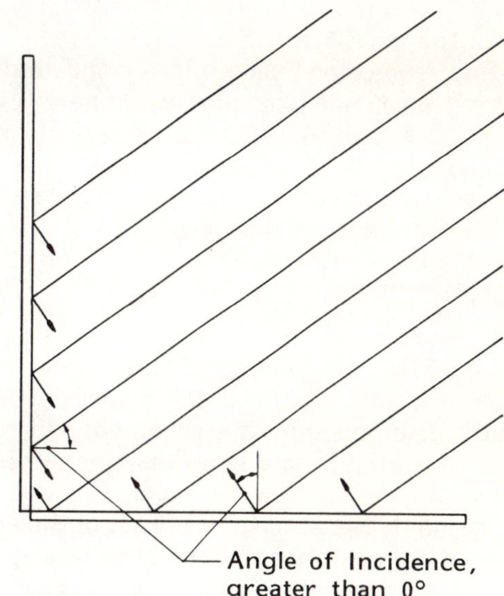

6b. Other surfaces.
Surface is seldom perpendicular to sun's rays. Less than maximum radiation is intercepted and more radiation is reflected.

Fig 6. Radiation on a sun-following surface.

dence, or angle between the sun's rays and a line drawn normal to the surface, Fig 5, is 0°. ("Normal to" means perpendicular to or at right angles—90°— to a surface.) Utilization of collector surface area is maximized and reflection off the collector is minimized, Fig 6.

Sun-following surfaces must pivot both horizontally and vertically to track the sun's changes in altitude and azimuth. Tracking requires a device to sense the sun's position, a mechanism to move the collector, and a fairly complex support structure. Tracking increases collector costs and maintenance requirements and is seldom cost-effective for agricultural solar collectors.

A support and control system to track the sun in only one direction—either altitude or azimuth—is much simpler. Fixed collectors that do not track the sun are even simpler and less expensive. Less solar energy is available to fixed collectors than tracking ones, but their cost is usually low enough that you can afford extra collector area (perhaps as much as 50%) to make up the difference.

In the northern hemisphere, fixed solar collectors generally receive the most solar energy per day if they face due south. But deviations up to 20° east or west of due south have little effect on daily solar energy collection. Deviations greater than 30° from south greatly reduce the amount of solar radiation received. Sometimes, southeast or southwest collector positioning may be desirable, such as when an obstruction blocks the morning or afternoon sun or when more heat is needed in the morning or afternoon or in areas with predominantly cloudy mornings or afternoons.

For maximum solar energy collection with a fixed collector, set the tilt angle so the angle of incidence is near 0° at solar noon at the time of year when you want to collect the most heat. The collector will perform almost as well as a sun-following collector in the middle of the day, but in the early morning and late afternoon when the angle of incidence is large, much of the solar radiation will reflect off the fixed collector surface.

Because the sun's altitude at solar noon varies with latitude and time of year, the appropriate tilt angle for a fixed collector depends on when and where it will be used. Fig 4 shows the solar radiation received by a sun-following surface and several fixed, south-facing surfaces at 40° north latitude: horizontal (0° tilt angle); 4/12—such as the south slope of a building roof (18.4°); tilt angle equal to the latitude (40° in this case); latitude plus 15° (55°); and vertical—such as a south building wall (90°).

Of the fixed surfaces, the one with a tilt angle equal to the latitude has the most consistent energy reception throughout the year. For applications requiring heat year-round, choose a collector with the tilt angle equal to latitude. The surface with a latitude plus 15° tilt angle receives the most energy during the coldest months of the year and is the best surface for collectors needed most during the winter heating season.

Horizontal surfaces receive the most energy in summer and the least in winter. The 4/12 slope performs much like a horizontal surface. In the northern U.S., the 4/12 slope is usually not a very good surface for meeting winter heating needs—it collects too much energy in summer and too little in winter. Also, dust, frost, and snow accumulations on 4/12 slopes can greatly reduce energy collection.

Vertical surfaces receive about 10% less solar energy than latitude plus 15° surfaces in the coldest months, receive very little solar energy in summer months, and have no problem with snow accumulation. Vertical surfaces can be supported on an existing wall. These factors make vertical surfaces a frequent choice for solar collectors used for winter heating in the northern U.S.

The final choice of fixed collector tilt angle depends on when you need the most solar energy, where you have space available for the collector (on the south wall, on the roof, or along the south side of the building), and which collector surface is least costly to build per unit of energy returned.

Atmospheric Effects

Even at noon on sunny days, not all the solar energy reaching the outside of the earth's atmosphere is available on the earth's surface. At 40° north latitude on a sunny March 21, only about 307 Btu/hr-ft² is available to a sun-following surface on earth, Fig 2, compared with about 428 Btu/hr-ft² available outside the atmosphere. Part of the energy arriving from the sun is reflected back into space at the top of the atmosphere, Fig 7. Some is absorbed by ozone, water vapor, carbon dioxide, and other compounds in the atmosphere. Another portion of the solar radiation is scattered by dust particles and water vapor. The amount of solar radiation scattered and absorbed depends on the length of the travel path through the atmosphere and the concentration of water vapor, carbon dioxide, dust, smog, etc., in the atmosphere.

When the sun is high in the sky, the length of the solar radiation's travel path through the atmosphere is fairly short. When the sun is low in the sky in early morning, late afternoon, and in the winter, the travel path is longer and more of the solar radiation is scattered and absorbed with less reaching solar collectors. Because of this, you might expect the noontime solar radiation to be greatest on June 21 when the sun is highest in the sky. Fig 2 shows that the solar radiation is actually greater at noon on Dec. 21.

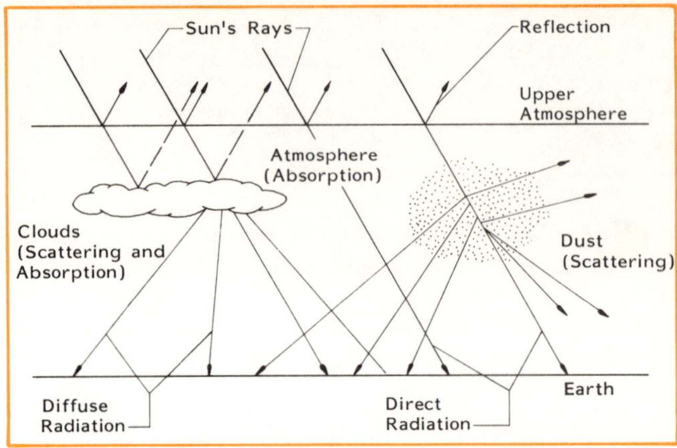

Fig 7. Atmospheric effects on solar radiation.

The solar radiation is greater in December because the earth is closer to the sun then, Fig 1. The earth is a little farther away from the sun in March than in December, but the shorter travel path for solar radiation through the atmosphere gives rise to a greater noontime solar intensity in March.

Solar radiation that passes directly from the sun to a collector surface without deflection is called direct or beam radiation. The scattered fraction is called diffuse radiation. The amount of diffuse radiation increases with the concentration of scattering materials in the atmosphere. On clear days, solar radiation is often about 85% direct and 15% diffuse. On overcast days, only diffuse solar radiation reaches the collector.

The quantity of solar radiation available on clear days is mainly a function of latitude (distance from the equator) and time of year, and is fairly predictable. Table 26, appendix, lists the clear day solar energy available at each hour of the day on the 21st of each month for several latitudes. Locations at the same latitude receive about the same amount of solar energy on clear days, except when there are large differences in elevation above sea level and/or average atmospheric conditions.

Clear day insolation values in Table 26 include both direct and diffuse radiation, but not radiation reflected off the ground or surrounding surfaces. The values are for average clear days. Actual insolation taken on any given clear 21st of the month could differ from the table value by 15%.

On the average, the amount of solar energy available to a solar collector is not as great as that shown in Table 26, because not every day is clear. Clouds scatter and absorb solar radiation and greatly reduce the amount available beneath them. Cloud cover varies from one location to another and varies greatly from one year to the next at a given location. Because cloud cover varies so much and is so hard to predict, the best we can do is average the solar radiation received at a location over a number of years and assume that, on the average, the same amount will be received in the future. Table 27, appendix, lists the average solar radiation received each month at various weather stations across the country. The insolation values include direct and diffuse radiation and radiation reflected off the ground, with average snow cover for each month taken into account. (Snow reflects more solar radiation than grass or soil.)

Data for Columbus, OH (40° north latitude), illustrate the effect of cloud cover on solar energy availability. On a clear December day, the expected insolation on a vertical collector is 1644 Btu/day-ft², Table 26. But on the average, only 837 Btu/day-ft² is available on a vertical collector in Columbus in December, Table 27. However, Lincoln, NE, which is also located at about 40° north latitude but has less cloudy weather, receives an average of 1275 Btu/day-ft² on a vertical collector in December, Table 27.

Tables 26 and 27, appendix, list the solar energy available **to** a solar collector and do not indicate the amount of heat energy actually available **from** a solar collector. To estimate heat output, you need to multiply the solar energy available at the collector surface by the collector efficiency. (See Efficiency section.) Values from Table 26 allow calculation of the maximum collector heat output. The peak heat output and temperature rise from a solar collector will occur shortly after solar noon on sunny days unless concrete or other massive materials are used in the collector. Then, peak temperature rise will occur later in the day due to storage effects. Values from Table 27 allow calculation of average collector heat output over cloudy and sunny days or the total heat available over a selected season.

Collector Shading

The heat output from a solar collector is reduced by shading. The extent of the reduction depends on when and how long the shading occurs, how much of the collector surface is shaded, and the nature of the obstruction causing the shading.

Although total day length is usually greater than six hours, most of the usable solar energy is received in about a 6-hr period from 9 a.m. to 3 p.m. (solar time). Collector shading that occurs outside this time interval will have little effect on overall collector performance. Any shading between 9 a.m. and 3 p.m. has a much greater effect. The longer the shading period is and the closer to noon it occurs, the greater the reduction in performance.

Shading by surrounding objects is most likely to occur in December and January when the sun is lowest in the sky (the lowest point is reached on Dec. 21). This is also the coldest time of the year, when maximum solar collector area is required to meet heating needs and shading is least tolerable. The higher a solar collector is above ground, the less likely it will be shaded.

The worst shading problems are caused by long, east-west buildings located south of the collector. Short buildings, single trees, silos, and obstructions southeast or southwest of a collector can also cause shading, but usually for shorter time periods. Winter shade from deciduous trees (trees that shed their leaves in fall) reduces collector performance 10%-40%, but reduced performance might be more acceptable than removing the trees.

Separation distance required to prevent solar collector shading depends on month, time of day, and height of the obstruction. Calculate separation distance by multiplying the solar angle factor from Table 1 by the obstruction height. Measure the calculated distance straight north from an imaginary east-west line drawn under the obstruction's highest point. The solar angle factors in Table 1 were chosen to prevent shading between 9 a.m. and 3 p.m. solar time.

Table 1. Solar angle factors (SAF).

Date	Latitude, degrees north			
	24	32	40	48
1a. For winter, SAF = cos(9 a.m. azimuth)/tan (9 a.m. altitude). See Fig 8.				
To prevent winter shading:				
	- - - - - -SAF- - - - - -			
Dec 21	1.5	2.0	3.0	5.4
Jan 21 and Nov 21	1.2	1.7	2.4	3.8
Feb 21 and Oct 21	0.8	1.0	1.4	1.9
Mar 21 and Sept 21	0.4	0.6	0.8	1.1
1b. For summer, SAF = 1/tan (noon altitude). See Fig 9.				
To produce summer shading:				
Apr 21 and Aug 21	0.2	0.4	0.5	0.7
May 21 and July 21	0.1	0.2	0.4	0.5
June 21	0.0	0.1	0.3	0.5

Example 1:

Find the separation required to prevent shading of a collector at ground level by a long, 20' high, east-west building between 9 a.m. and 3 p.m. during the winter heating season at 40° north latitude, Fig 8.

Answer:

Maximum separation is required on Dec. 21 when the sun is lowest in the sky. From Table 1a, the solar angle factor, SAF, for Dec. 21, 40° north latitude, equals 3.0.

$$d = \text{SAF} \times h$$
$$= 3.0 \times 20' = \mathbf{60'}$$

d = separation distance, ft
SAF = solar angle factor
h = obstruction height, ft

Example 2:

Find the length of overhang needed to completely shade an 8' high vertical collector at noon, May through July, at 40° north latitude, Fig 9.

Answer:

Use the solar angle factors in Table 1b to calculate required overhang lengths. The shortest overhangs are required in June when the sun is highest in the sky. To shade a collector during May, June, and July, you need a longer over-

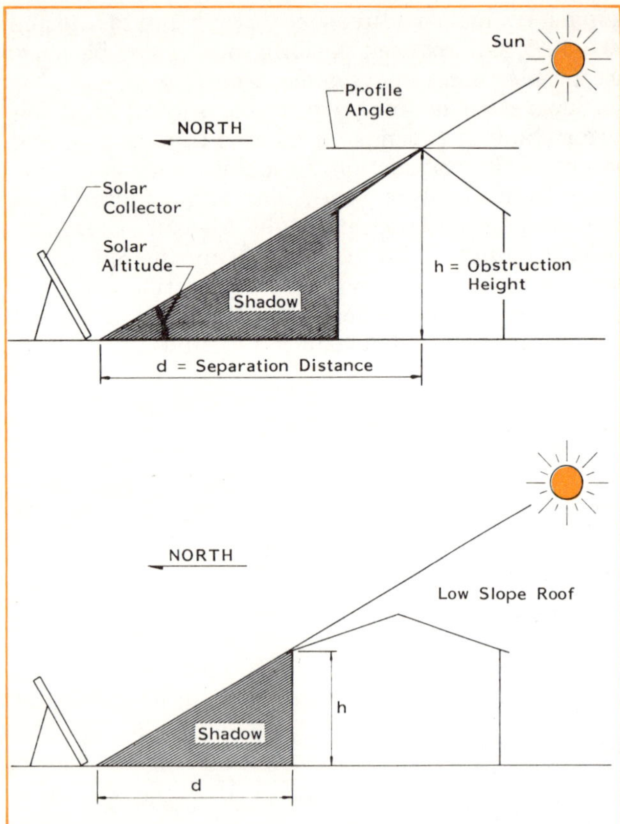

Fig 8. Separation distance to prevent shading.

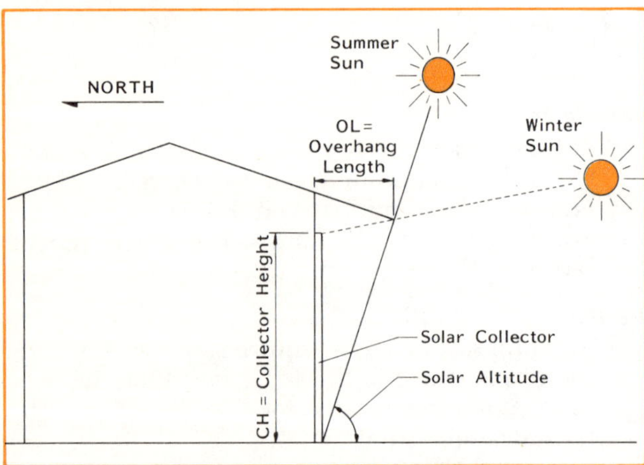

Fig 9. Overhang length to shade collector.

hang. Find the solar angle factor for May and July, 40° north latitude, from Table 1.

$$SAF = 0.4$$
$$OL = SAF \times CH$$
$$= 0.4 \times 8' = 3.2'$$

OL = overhang length, ft
SAF = solar angle factor
CH = collector height, ft

A 3.2′ overhang will completely shade the collector at noon May 21 through July 21, but will also cause some shading at noon during the heating season. It is probably better to use a shorter overhang (e.g., 2′) to reduce winter shading.

Solar Collectors

Basics

The purposes of the solar collectors discussed in this handbook are to:
- Intercept radiation from the sun.
- Convert this solar energy into thermal (heat) energy.
- Transfer the heat energy to a fluid (air or liquid) that carries it to the point of use or heat storage.

Most solar collectors include the following components: an absorber; a glazing or cover; a support structure; and a heat transfer fluid. The absorber is heated by the sun and transfers the heat to the fluid moving over or through it. The glazing or cover plate transmits the solar radiation to the absorber, and reduces heat losses. The support structure is the mount for the collector components. The heat transfer fluid moves the heat from the collector to storage or use.

Basic types of solar collectors are flat plate and concentrating. A flat plate collector with reflectors has some characteristics of both, Fig 10.

Flat Plate Collectors

An essential part of a flat plate collector is the absorber. It absorbs solar energy, heats up, and then transfers the heat energy to a fluid moving over or through it. It is really a device that converts solar radiation into heat energy.

A number of area terms are used to describe flat plate collectors. **Gross collector area** is the surface area calculated from outside dimensions (e.g. a 4′x8′ collector has a gross area of 32 ft²). In this book, gross collector area is considered to be the **energy intercepting area** and is the area used in efficiency and heat collection equations. In flat plate collectors, energy intercepting area equals absorbing area, while in concentrating collectors, intercepting area exceeds absorbing area.

Net collector area is the gross area minus the area of cover supports and framing members that shade the absorber. Many air-type flat plate collectors use corrugated, crimped, or finned absorbers to improve heat transfer from the absorber to the airstream. Such absorbers may have a material surface area greater than the gross collector area and may improve collector efficiency, but they do not increase the energy intercepting area used in calculations.

Most flat plate collectors are fixed (do not track the sun). They collect both direct and diffuse radiation, so they may produce small amounts of heat even on overcast days when all solar radiation is diffuse.

In air-type flat plate collectors, a moving air stream picks up heat from one or both sides of the absorber. In the simplest type, bare plate, an air channel is formed behind the absorber with back and side plates. The air moves through the channel and picks up heat conducted through the absorber, Fig 11a. A covered plate collector, Fig 11b, is usually more efficient. A transparent cover over the absorber forms the air channel. Air passes between the absorber and the cover to pick up heat. The transparent cover, or glazing, reduces heat losses from the absorber. A suspended plate version is even more efficient. The absorber is between a transparent cover and a back-plate, and air is directed either behind or on both sides of it, Figs 11c and 11d.

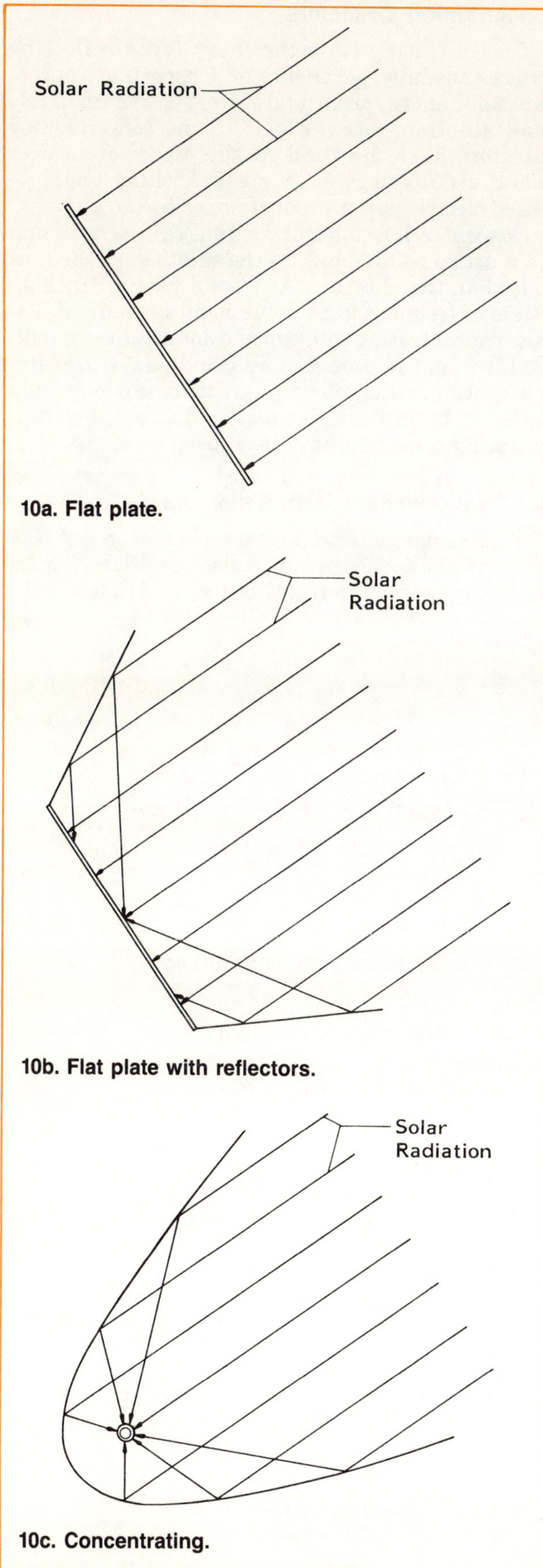

Fig 10. Basic types of solar collectors.

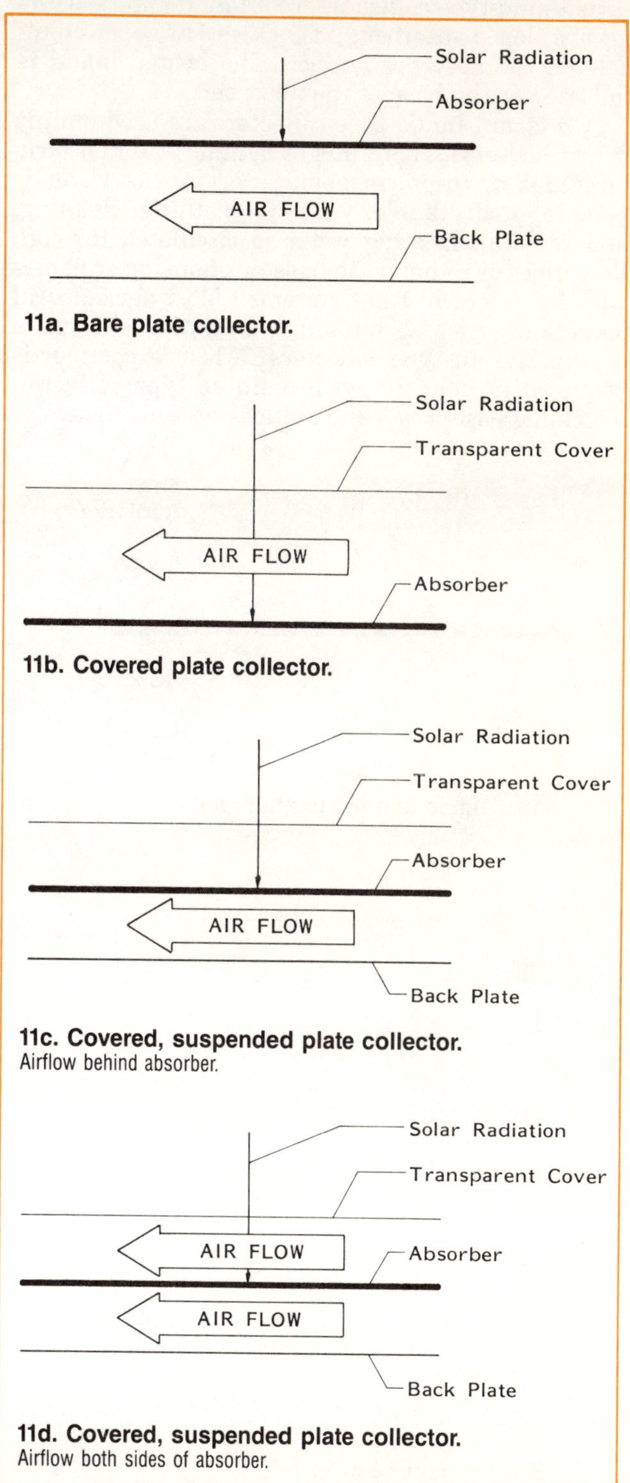

Fig 11. Air-type flat plate solar collectors.
Cross sections of collectors.

In liquid-type collectors, a liquid (usually water or water plus antifreeze) removes heat from the absorber. In many liquid-type flat plate collectors, the liquid flows in tubes bonded to or part of the absorber, Fig 12. The tubes are in a serpentine pattern, Fig 12c, or a parallel pattern with headers or manifolds on each end, Fig 12d. Covered plate liquid-type collectors are much more common than bare plate ones. In some liquid-type collectors, cool liquid enters at the top of a sloped absorber and trickles down open channels on the absorber surface. The heated liquid is collected at the base of the absorber.

On farms, liquid-type collectors are used mainly to provide service hot water or heating water for farm buildings or the farm home. (Service hot water is potable, or drinkable, water for bathing, cleaning, and cooking. Heating water is circulated through floor pipes or radiators in livestock housing or homes with hot water heating systems.) Most agricultural operations, such as building ventilation and grain drying, use air-type collectors. A heat exchanger is required to heat air with a liquid-type collector, which increases costs and reduces system efficiency.

Concentrating Collectors

Concentrating collectors have large reflecting surfaces (usually parabolic) or lenses that concentrate solar energy from a large area onto a relatively small absorbing area, Fig 13. Some concentrating collectors heat the fluid in the absorber to over 2000 F, usually to generate steam for industrial processes, electric power production, or to run engines.

Concentrating collectors use only direct radiation and must be pointed toward the sun to keep the rays focused on the absorber. At least a partial tracking system or frequent focus adjustment is required. Because concentrating collectors do not use diffuse radiation, no heat is produced on cloudy days. Because most concentrating collectors are more expensive and harder to build than flat plate collectors, few concentrating collectors have been used on farms.

Flat Plate Collectors with Reflectors

These solar collectors have some features of both flat plate and concentrating collectors. Reflectors increase the energy intercepting area and concentrate,

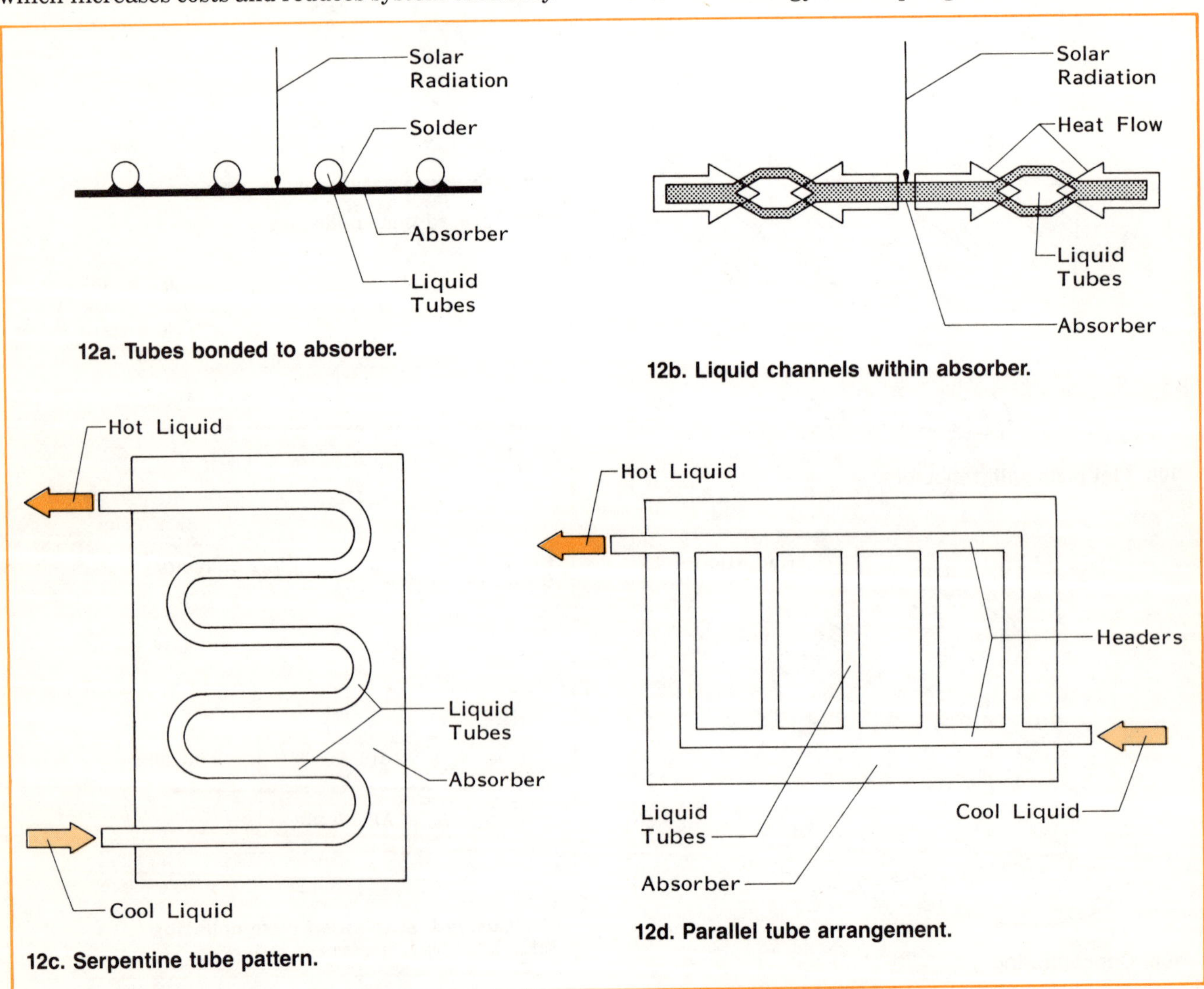

12a. Tubes bonded to absorber.

12b. Liquid channels within absorber.

12c. Serpentine tube pattern.

12d. Parallel tube arrangement.

Fig 12. Liquid-type flat plate collectors.

to an extent, the energy available to a flat plate collector. With reflectors, temperatures in a flat plate collector can be increased. However, reflectors can use only direct radiation and work best in sunny regions.

Reflectors include plywood wings painted white or lined with aluminum foil, concrete slabs painted white, and parabolic sheets lined with aluminized film, Fig 14. Consider the value of the extra energy collected over the life of the reflectors against the cost of the materials and labor required to build and maintain them. Fresh snow reflects solar radiation fairly well and slightly increases the energy available to flat plate collectors at no cost to you.

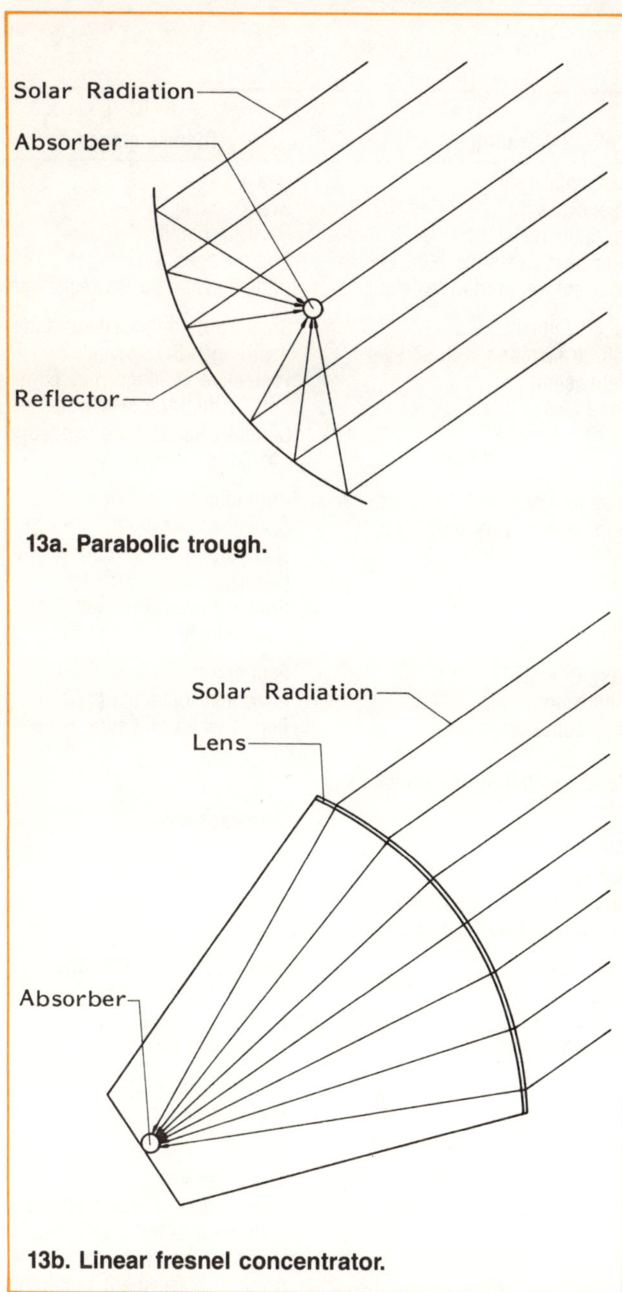

13a. Parabolic trough.

13b. Linear fresnel concentrator.

Fig 13. Types of concentrating solar collectors.

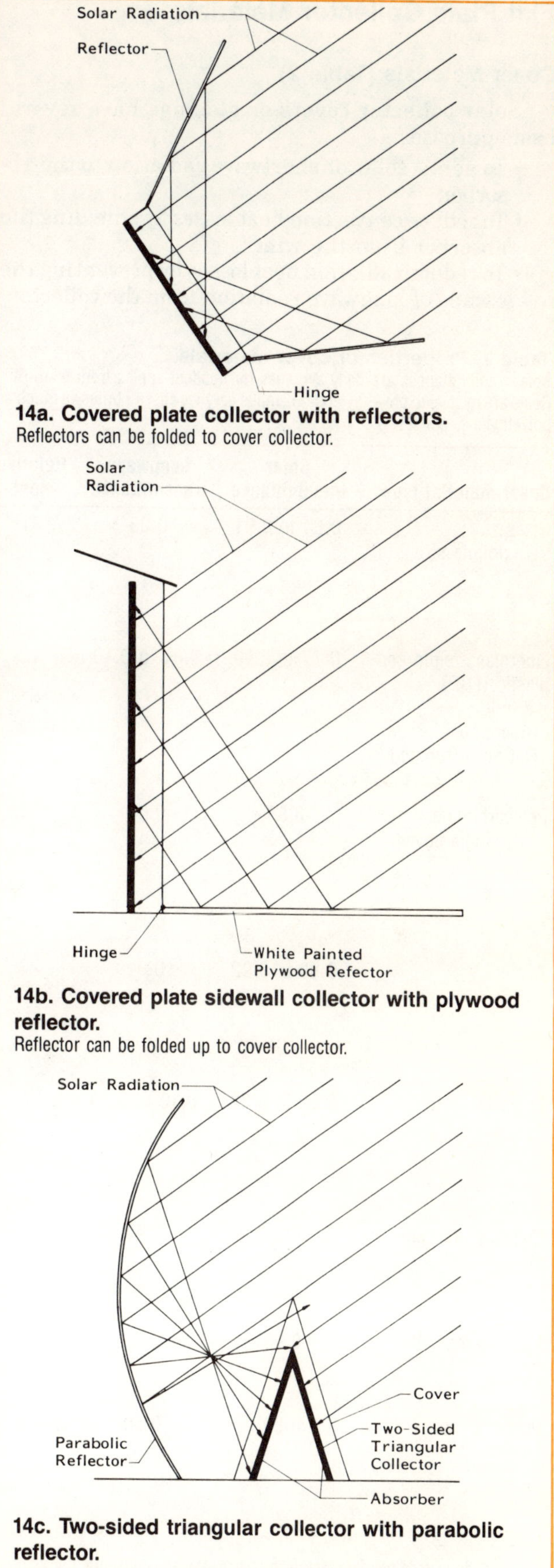

14a. Covered plate collector with reflectors.
Reflectors can be folded to cover collector.

14b. Covered plate sidewall collector with plywood reflector.
Reflector can be folded up to cover collector.

14c. Two-sided triangular collector with parabolic reflector.

Fig 14. Flat plate collectors with reflectors.

Flat Plate Collector Materials

Cover Materials (Table 2)

Solar collector covers or glazings have several basic purposes:

- To admit solar or shortwave radiation to the absorber.
- To reduce convection heat losses by shielding the absorber from the wind.
- To reduce radiation heat losses by preventing the escape of longwave radiation from the collector.

All objects that are warmer than their surroundings lose energy by radiation. Very hot objects, like the sun, radiate mostly high energy waves with short wavelengths (shortwave radiation). Objects at expected temperatures and only slightly warmer than surrounding temperatures, such as absorber plates in solar collectors, radiate or emit lower energy waves with long wavelengths.

When an absorber is not covered (as in bare plate collectors), much of the absorbed energy is lost to the wind. When an absorber is covered (as in covered plate collectors), the losses are reduced. A good cover

Table 2. Properties of cover materials.
Solar transmittances are daily averages for incident angles from 0° to 67°. Costs are relative to polyethylene. Example: acrylics cost 43 times as much as polyethylene, per ft^2.

Cover material type*	Solar transmittance	Longwave transmittance	Relative cost	Advantages	Disadvantages
Glass ⅛" double strength	0.84 to 0.91	0.03	21	Neat appearance Cleans easily Abrasion resistance High heat tolerance (400 F) Excellent weathering resistance	Heavy Breaks easily Hard to install Sharp edges Poor thermal stress resistance
Fiberglass reinforced plastic (FRP) 40-mil Greenhouse grade Flat or corrugated	0.77 to 0.90	0.01 to 0.10	12	Easy to install Can be fastened with screws Lightweight Good availability Tough, durable	Solar transmittance deteriorates with age (5-15 yr life) Protective coatings may come off or degrade over time Cannot tolerate long exposure to temps over 200 F
Polycarbonate ⅛", single cover	0.88	0.03	21	Easy to install Lightweight, very tough	Combustible Scratches easily High expansion & contraction Becomes brittle with age Solar transmittance deteriorates Low heat tolerance (170 F)
Acrylics	0.80 to 0.92	0.01	43	Easy to install Lightweight Nice appearance Long life (25 yr) Good weathering resistance	Expensive Low heat tolerance (170 F) Bends when one side heated
Polyester films	0.80 to 0.89	0.10 to 0.20	7	Low cost Easy to install Strong Fairly long life (10-15 yr) Fair weathering resistance	Scratches easily
Polyvinyl fluoride film	0.90 to 0.94	0.45	9	Light weight Very strong Excellent weathering resistance Scratch resistant	Becomes brittle with age Shrinks upon heating Low heat tolerance (225 F) Hard to install
Fluorocarbon film	0.96	0.58	12	Resistant to deterioration Good inner glazing High heat tolerance (400 F)	Stretches with age (5-15 yr) High longwave transmittance
Polyethylene 4-mil	0.89	0.80	1	Tough Lightweight	Punctures easily Susceptible to wind damage when collector is deflated Solar transmittance deteriorates Annual replacement required

*Formulations used by manufacturers vary greatly and consequently, so do material properties. Use this table only as a general guide. See manufacturer's literature for more accurate numbers.

material has a high transmittance to shortwave radiation and a low transmittance to longwave radiation.

Cover materials that allow shortwaves from the sun to pass but block longwaves from the absorber, cause the "greenhouse effect." The absorber is warmed by solar energy and the heat is trapped within the collector. An ideal cover material has a shortwave or solar transmittance of 1.0 (100% of the solar radiation passes through the coverplate) and a longwave transmittance of 0 (none of the longwave radiation escapes). But, with real materials, part of the incident solar radiation is reflected and part is absorbed by the cover, and part of the longwave radiation escapes. Short- and longwave transmittances for a number of materials are listed in Table 2.

The transmittance of a material is not constant. It varies from a maximum at a 0° incident angle (radiation normal to surface) to 0 at a 90° incident angle (radiation parallel to surface), Table 3. Less radiation is transmitted through a material at large incident angles because more is reflected off the surface—like a rock skips off a water surface. The transmittance decreases rapidly at incident angles larger than 60°. Because the solar angle of incidence on fixed, flat plate collectors usually exceeds 60° before midmorning and after midafternoon, most of the energy is collected by fixed collectors during the middle part of the day.

Adding cover layers reduces energy losses from a collector, but also reduces the radiation transmitted to the absorber. Each additional layer absorbs and reflects another fraction of the solar radiation.

In addition to good transmittance, other desirable properties of collector cover materials are:
- Resistance to deterioration by ultraviolet radiation, moisture, atmospheric pollutants (weathering resistance), and wide temperature ranges.
- Little expansion and contraction with temperature changes, or a rate of expansion that matches that of the framework holding the cover.
- High impact strength against hailstones, rocks, birds, and livestock.
- Low flammability.
- Economy. An expensive but durable material may be more economical than a cheaper one requiring annual replacement.
- Ease of installation and maintenance.
- Resistance to wind damage and abrasion at supports.
- Resistance to deterioration at high temperatures. When the fluid is stagnant and the sun is shining, temperatures can exceed 300 F.
- Lightweight.
- Resistance to static electricity charges which attract dust (reducing solar transmittance) and make the material difficult to handle.
- Ease of sealing against leakage.

Glass is most common on collectors with recirculated heat transfer fluids. (See Solar Systems.) It has excellent transmittance that does not change over time. It is rigid and has good abrasion resistance, so it does not sag or wear through at supports. Single-strength (0.085″ to 0.100″ thick), double-strength (0.115″ to 0.133″ thick), tempered (high-strength), and low-iron glass are available. Low-iron glass has a higher solar transmittance than other glass. The edge of low-iron glass appears water-white rather than the usual greenish-blue.

Table 3. Solar transmittance through double-strength window glass.

Incident angle degrees	Solar transmittance	
	One layer	Two layers
0-20	0.87	0.77
30	0.87	0.76
40	0.86	0.75
50	0.84	0.73
60	0.79	0.67
70	0.68	0.53
80	0.42	0.25
90	0.00	0.00

The biggest disadvantage of glass is brittleness. Rocks, hail, and even stresses caused by thermal expansion and contraction may break it. Vertical mounting or a wire-screen cover reduce hail damage but also reduce the amount of solar energy reaching the collector. The thermal expansion of glass is less than that of wood or steel. Glass attached directly to a wood or steel frame may break when subjected to temperature extremes. Use expansion gaskets between the glass and frame.

A number of transparent plastics can be used as collector covers. One of these, polyethylene, has a high initial solar transmittance and very low first cost, and was used on many early collectors. But it traps little longwave radiation and usually requires annual replacement due to ultraviolet degradation. More durable materials are probably more cost-effective in the long run.

Fiberglass reinforced plastics (FRP) are popular cover materials for low temperature agricultural collectors. The type used for collectors is sometimes called "greenhouse grade fiberglass." Its solar transmittance is less than glass but much better than skylight material. It is clear and almost see-through.

Not all materials sold as "greenhouse grade fiberglass" possess all the characteristics desirable in a cover material. The FRP suitable for solar collectors is **not** skylight and patio cover material.

Because FRP is tougher than glass and expands at nearly the same rate as wood and steel, it is easier to install and can be connected directly to a collector frame. However, even with FRP, keeping joints sealed is difficult. FRP is more flexible than glass and can be fastened to curved surfaces, but it requires more support. Corrugated FRP is more rigid than flat FRP and is harder to seal against air leaks. Use caulked, corrugated wood strips or closed cell foam strips for sealing.

Heat, ultraviolet radiation, and moisture cause FRP to gradually deteriorate. Some FRP is guaranteed for 15 to 20 years on greenhouses, but not yet on solar collectors where temperatures can be much higher. FRP develops fiber bloom as the binding plastic resin breaks down, exposing the glass fibers. Dirt and fungi trapped among these fibers reduce solar transmittance. FRP can be treated or coated to prevent fiber bloom, but the coating may peel off or degrade over time and require reapplication. Some plastics are ultraviolet resistant.

Other cover materials and their advantages and disadvantages are listed in Table 2.

Absorber Coating Materials

A good absorber in a solar collector:
- Absorbs a high percentage of incoming solar radiation.
- Loses minimum energy to the collector's surroundings.
- Efficiently transfers absorbed energy to the collector fluid.

All solar radiation striking an opaque (nontransparent) surface is either reflected or absorbed. The fraction that is absorbed (always a decimal between 0 and 1) is called the absorptance. The reflected fraction, the reflectance, is 1 minus the absorptance. A good absorber has an absorptance near 1.

Dark surfaces have high absorptances—so paint collector absorbers black if they are not naturally dark. Flat black paints reflect less energy than glossy ones. Any flat black paint that can withstand temperatures up to 300 F without cracking, peeling, or breaking down can be used. Use an appropriate primer or surface treatment before painting.

Let the paint dry completely before installing the cover. Some paints give off vapors (called outgassing) as they dry, which can condense on the collector cover and reduce transmittance. Also, the vapors from some paints (particularly latex) have unpleasant odors—an important consideration for collectors on shops or homes.

To avoid painting and paint adherence problems, you can buy metal that already has a baked enamel, flat, very dark surface.

Absorptances for a number of common surfaces are shown in Table 4. The solar absorptance of a surface varies with angle of incidence, just as transmittance does. At high angles of incidence a greater percentage of the insolation is reflected, Table 5.

Most surfaces with high solar absorptance also have a high longwave emittance. The longwave emittance of a surface (always a decimal between 0 and 1) is its tendency to radiate longwave energy. A solar absorber heats to a temperature above its surroundings and then radiates or emits energy as longwave radiation. An ideal absorber has a longwave emittance near 0.

"Selective surfaces" have both high solar absorptance and low longwave emittance. Many are special factory-applied coatings and are not suitable for field applications. Selective paints are available, but the cost per gallon is high. Selective surfaces can operate at higher temperatures because they lose less energy by radiation. Properties of several commercial, selective coatings are given in Table 6. Select one guaranteed to withstand high temperature exposure.

Well-weathered galvanized metal with a rough, dull gray surface is a natural mildly selective surface, Table 4. Painting the metal flat black increases the solar absorptance, but destroys the selective surface. In some cases, the unpainted surface is probably more cost-effective.

Table 4. Absorptance and emittance of common surfaces.
Normal solar incidence (0° incident angle).

Material	Solar absorptance	Longwave emittance
Flat very dark paint	0.95-0.99	0.95-0.99
Dark concrete and stone	0.65-0.80	0.85-0.95
Colored paints; brick	0.50-0.70	0.85-0.95
Bright aluminum paint	0.30-0.50	0.40-0.60
Dull metals: copper, brass, aluminum	0.40-0.65	0.20-0.30
Weathered galvanized steel	0.80	0.28
White paint	0.23-0.49	0.92

When preparing a **new** galvanized metal surface for use as an absorber plate, wear rubber gloves to protect yourself from the chemicals and the metal from skin oils. Clean the metal with a solvent, such as acetone, to remove manufacturing oils, and then wash the surface with detergent. Etch the metal with vinegar or a weak acid (e.g. 6:1 water/muriatic acid mixture) and rinse. Apply a galvanized metal primer and two coats of flat black paint.

Table 5. Solar absorptance of flat black paint.

Incident angle degrees	Solar absorptance
0-20	0.96
30	0.95
40	0.94
50	0.92
60	0.88
70	0.82
80	0.67
90	0.00

The solar collector absorber must transfer its absorbed energy to the heat transfer fluid. In liquid-type collectors, heat moves by conduction through the flat absorber plate to the tubes containing the liquid. Both the absorber and tube materials need high thermal conductivities, and the bond between the tubes and plate must provide a good thermal path. Metal absorber plates are best, and the most common in order of increasing conductivity and cost are steel, aluminum, and copper. The metals are painted black or coated with a selective surface material. Higher absorber conductivity permits greater tube spacing. Plastic and rubber absorbers are also available. Because of the low conductivity of plastic and rubber, the

Table 6. Absorptance and emittance of selective surfaces.
From *Solar Heating Materials Handbook* by Homann, Hilleary, and Darnall.

Absorber coating	Solar absorptance	Longwave emittance
Aluminum oxide and molybdenum dioxide	0.9	0.1-0.4
Black chrome (Chromium metal and chromium oxide on nickel-plated metals, copper, or steel)	0.91-0.96	0.07-0.16
Black copper (Cupric oxide and cuprous oxide on copper, nickel, or aluminum)	0.81-0.93	0.11-0.17
Black nickel (Nickel oxide on nickel, iron, or steel)	0.89-0.96	0.07-0.17
Iron oxide on iron or steel	0.85	0.08
Lead oxide	0.98-0.99	0.22-0.40
Stainless steel oxide on stainless steel	0.89	0.07
Selective paint	0.90	0.30

tubes must be very close together to obtain good efficiency.

The best bonds between metal tubes and absorber plates are solid beads of weld or solder or a thermal bonding mastic. Use only silver solder inside the collector itself. A 50:50 tin/lead solder is acceptable for the rest of the plumbing. Clamps, wires, press fits, and spot welds do not provide as good a thermal conducting joint. Bonding is avoided in tube-in-sheet absorbers—the liquid tubes are part of the absorber plate, Fig 12.

Chemical reactions between dissimilar metals can be a serious problem, especially with copper and aluminum or galvanized steel. Consider corrosion problems before installing a liquid solar system.

In air-type solar collectors, the heat transfer fluid passes over the entire absorber and picks up heat directly from the surface. In bare and suspended plate collectors, heat must be conducted to the back side of the absorber plate, so it should be made of metal. Absorber plates for other collector types can be of any convenient material: glass, plastic, concrete, or high temperature insulation board. The material must withstand sunlight and collector stagnation temperatures without sagging, deteriorating, melting, or burning. Paint the surface black if it isn't naturally dark.

Cautions about using wood and/or fiberglass in solar collectors:
- Extended exposure of wood to temperatures above 200 F can reduce its strength and ignition temperature.
- Wood exposed to temperatures above about 400 F may spontaneously ignite.
- A proposed UL standard suggests 194 F as the maximum temperature for wood in collectors.
- Exposing fiberglass cover sheets to temperatures above 200 F accelerates degradation.
- Cover, ventilate, or shade solar collectors that are not used during summer months.

Absorber-to-air heat transfer depends on flowrate per unit area, turbulence, and absorber material surface area. Up to a point, increasing air velocity improves heat transfer. (See Collector Design.) Increasing the absorber surface area also improves heat transfer and often improves collector efficiency by decreasing reflective losses. Corrugated, crimped, and finned absorbers, which provide greater surface area than simple flat surfaces, have all been tested, Fig 15. With typical metal roofing and siding sheets, airflow direction over corrugated surfaces (parallel vs. perpendicular) makes little difference in collector performance. Make airflow parallel to the deeper corrugations of crimped and finned absorbers.

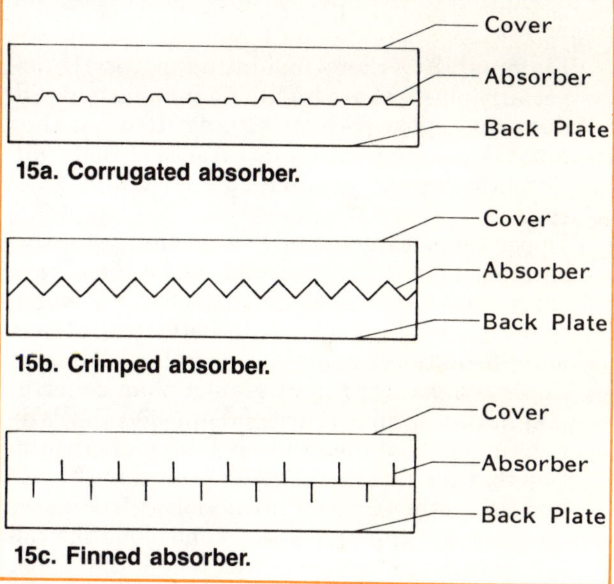

15a. Corrugated absorber.

15b. Crimped absorber.

15c. Finned absorber.

Fig 15. Air-type suspended plate solar collectors.
Airflow toward the viewer.

Insulation

(See Table 13 in the Insulation section and Table 7.)

In addition to losing heat by radiation from the absorber, solar collectors lose heat by conduction through the side and back plates. Insulate collectors to reduce conduction heat losses. Collectors that preheat ventilation air for livestock buildings operate at temperatures only slightly higher than outdoor temperatures and need to be insulated to about R=6. Solar collectors that heat recirculated air or water usually operate at temperatures much higher than outdoor temperatures and need to be insulated to about R=11. Insulate ducts and pipes passing through unheated spaces to the same R-value used in the collector.

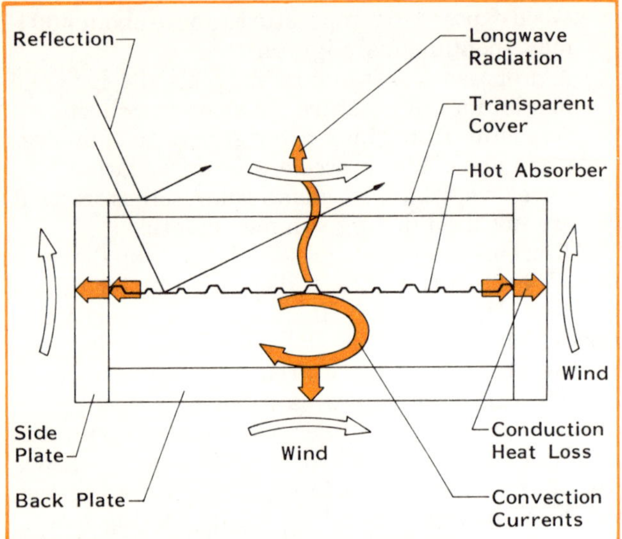

Fig 16. Heat losses from a solar collector.

Collectors mounted directly on insulated building walls may not need additional insulation.

In addition to R-value, consider these insulation properties:

- Flammability. Some insulating materials (especially plastics) are highly flammable and produce toxic gases when burning. If used, they must be covered with a fire-resistant material. Consult your insurance company and local codes.
- Upper temperature limit. Heat tolerance of insulating materials varies, but all of them have some upper temperature limit beyond which they melt, burn, or chemically deteriorate. Make sure insulation used in a solar collector has an upper temperature limit greater than collector stagnation temperatures (about 300 F). Polystyrene melts at about 165 F. Fiberglass insulation that is held together with an organic binder and some other types of insulation give off gases (outgas) that can condense on and cloud the collector cover at temperatures below the upper limit. Use Table 7 as a guide, but consult the manufacturer for information on a specific insulation brand.
- Durability. If insulation will be exposed to harsh conditions—sunlight, moisture, rubbing or chewing by livestock, birds, and rodents—either protect the insulation or select a type that can withstand the conditions. Polystyrenes and some other plastic insulations decay in sunlight. All materials lose their insulating value when wet—closed cell plastic foams may be a good choice for wet environments.
- Cost. When comparing insulation costs, include first cost of the insulation and any protective coverings, installation labor, and life for installations with equal R-values.

Foil-faced glass fiber board is usually a good insulation inside or outside solar collectors.

Table 7. Insulation upper temperature limits.
Consult the manufacturer for information on specific products—some materials can emit poisonous gases when overheated.
From *Solar Heating Materials Handbook* and UL Standard. Most conservative values were used.

Insulation type	Outgassing begins (F)	Approximate upper temperature limits (F)
Glass fiber building insulation	302	350
Glass fiber board	212	450
Glass fiber		
with low binder content	302	850
with no binder	482	850
Polystyrene		165 (melts)
Polyurethane		200
Isocyanurate	122	250

Efficiency

Converting solar energy to heat energy is not 100% efficient. Tables 26 and 27, appendix, show the amount of solar energy available at a collector surface—not the amount of heat available from the collector. Efficiency quantifies how well a collector converts available solar energy into heat energy. Collector efficiency helps you choose among collector types or predict the amount of energy or the temperature rise from a collector.

In this handbook, collector efficiency is the heat energy output over a time period divided by the solar energy available on the gross collector surface over the same time period. Because of energy losses from the collector, efficiency is always less than 100%.

Eq 1.

$$EFF = 100 \times CE \div AE$$

- EFF = collector efficiency, %
- 100 = gives efficiency in % (For efficiency as a decimal, do not multiply by 100).
- CE = collected energy, Btu per unit of time
- AE = available energy, Btu per unit of time

Available solar energy is at the collector's tilt angle. Collector efficiency is for the time period over which the energy was collected. It is possible to calculate instantaneous, all-day, and average efficiency over a season for a collector. Because collector efficiency varies with time of day and weather conditions, these three efficiency values usually differ. Know which type of efficiency is being considered when a value is given. See Collector Sizing for examples using efficiency values.

Factors that affect collector efficiency include:

- Difference between the average collector fluid temperature and the outdoor temperature. The greater this difference, the greater the heat loss from the collector. Average fluid temperature approximately equals the average of inlet and outlet temperatures, i.e. (inlet temperature + outlet temperature) ÷ 2. Temperature rise is affected by the available solar energy and the fluid flowrate per unit of collector area. Collectors that heat outdoor air generally operate with relatively low temperature differences and high efficiencies.

- Fluid flowrate/unit area, which affects exposure of the fluid to the hot absorber and the cold glazing. High flowrates reduce exposure time, fluid temperature rise, and heat losses from the collector. High flowrates also increase fluid velocity and improve absorber-to-fluid heat transfer. So, up to a point, higher flowrates improve collector efficiency. Maximum collector efficiency is limited by the cover and absorber properties; once this efficiency is reached, higher flowrates cannot improve it. Also, pressure drop in the collector increases as flowrate increases. Pressure drop measures fluid friction or resistance to flow and requires extra fan or pump power. The selected flowrate must balance these factors. Air-type collectors heating livestock ventilating air are usually operated at 2 to 3 cfm/ft^2 of collector surface area and 500 to 1000 fpm air velocity past the absorber.
- Wind speed. Wind increases heat loss and reduces collector efficiency.
- Insulation reduces heat loss and increases efficiency. See Insulation.
- Length of fluid flow path through the collector. Up to a point, the longer the fluid is in the collector, the hotter it gets. As length is added to a collector, heat losses increase and in the extreme can equal the solar energy added by the extra area. If a collector is too long, the fluid temperature reaches its maximum part way through the collector and the remaining area is useless.
- Collector type and tilt angle. With the same insulation and fluid flowrates, air-type collectors (Fig 11) in order of increasing efficiency are:
1) Bare plate.
2) Covered plate, single channel.
3) Covered, suspended plate—airflow above the absorber.
4) Covered, suspended plate—airflow under the absorber.
5) Covered, suspended plate—airflow both sides of the absorber.
Collectors operating at a poor tilt angle (large angle of incidence) have poor efficiency, because much of the solar radiation reflects off the cover.
- Cover transmittance. High shortwave radiation transmittance and low longwave transmittance increase efficiency. See Cover Materials.
- Number of covers. Multiple covers reduce heat loss, but add cost and reduce solar transmittance. Most collectors in recirculating systems have two covers; most in single pass systems have one. See Solar Systems.
- Absorber properties. Increase efficiency with high solar absorptance, low longwave emittance (selective surfaces), and absorbers with corrugated, finned, or crimped surfaces. See Absorber Materials.

- Care in construction. Poorly-built collectors with many air leaks have very low efficiency. Collector components must fit together and be well sealed.

Maximum efficiency can be achieved only at high cost; multiple low-iron glass covers, selective absorber coatings, high insulation levels, etc. Select only those features needed to maintain reasonable efficiency in your application. Solar collectors that provide service hot water or space heat for homes must operate at low fluid flowrates and high temperature differences. Such collectors are relatively costly. Most agricultural collectors have higher fluid flowrates and lower temperature differences, so they can be simple and fairly inexpensive and still operate with good efficiency. Collector efficiencies are included on the collector examples in this handbook.

Solar **collector** efficiencies should not be confused with solar **system** efficiencies. Solar system efficiencies are based on energy supplied to the point of use and are always less than collector efficiencies because of losses in ducts, pipes, fans, pumps, and storage. See Energy Utilization.

Solar Energy Storage

Install heat storage to save, and in most agricultural installations, to effectively utilize solar energy for night or cloudy days. Energy is stored by passing solar-heated air or liquid through a heat storage material (active systems) or by allowing direct sunlight to fall on the material (passive systems). Energy is recovered by passing cool air or liquid through the warm storage material or by direct radiation from the material. Choose storage materials that can store a lot of energy without taking up too much space (high volumetric specific heat). See Table 8.

Table 8. Heat storage materials.
Specific heat is for sensible heat storage.
Glauber's salt melts at 91 F and has a latent heat of fusion of 108 Btu/lb.

Material	Density lb/ft^3	Specific heat Btu/lb-F	Volumetric specific heat Btu/ft^3-F
Water (8.33 lb/gal)	62.4	1.0	62.4
$^{50}/_{50}$ water/glycol (8.8 lb/gal)	65.8	0.8	52.6
Clay bricks	135.0	0.2	27.0
Sand	95.0	0.2	18.0
Rock (¾″-3″ diameter)	100.0	0.2	20.0
Concrete (solid)	150.0	0.2	30.0
Glauber's salt			
Solid	100.0	0.5	50.0
Liquid	70.0	0.8	56.0

Specific heat is the number of Btus required to raise the temperature of 1 lb of material 1 degree F. Volumetric specific heat is the number of Btus required to raise the temperature of 1 ft^3 of material 1 degree F. Storage material weight is an important consideration, because extra floor support—even under concrete slabs—may be required. Besides weight and volume, consider cost, availability, life of the material, and compatibility with your solar system.

Sensible Heat Storage

You can store solar energy as sensible heat in a large mass of water, rock, concrete, etc. (Sensible heat raises the temperature of a material.) The energy is regained as the material cools. To calculate the amount of energy stored:

Eq 2.

$$Qs = WT \times SH \times TD$$

or

Eq 3.

$$Qs = VOL \times SHv \times TD$$

Qs = quantity of sensible heat, Btu
WT = weight of storage material, lb
SH = specific heat of storage material, Btu/lb-F
SHv = volumetric specific heat, Btu/ft^3-F
VOL = volume of material, ft^3
TD = temperature difference, F

Water is the usual storage material in liquid-type solar systems. Use only softened or low mineral content water to prevent mineral deposits. Storage tanks must be leak free, corrosion resistant, and able to stand the heat and weight of the hot water. Steel, fiberglass reinforced plastic, and waterproofed concrete have been used. Glass or ceramic-lined tanks are common in service hot water systems. Steel tanks must be lined inside to prevent corrosion and must also be lined outside if they will be buried. Insulate tanks on all sides to at least R=11 to reduce heat loss to the air or the soil.

Example 3:

How much energy is stored when a 5'x5'x4' rectangular water storage tank is heated from 50 F to 120 F?

Answer:

From Table 8, specific heat of water is 1 Btu/lb-F.

WT = VOL x DEN
= 5x5x4 x 62.4
= 100 x 62.4
= 6240 lb
Qs = WT x SH x TD (Eq 2)
= 6240 x 1 x 70
= **436,800 Btu**
WT = weight of storage material, lb
VOL = volume of storage material, ft^3
DEN = density of storage material, lb/ft^3

With an air-to-water heat exchanger, heat from air-type collectors can be stored in water. Rock or concrete is generally the heat storage material in air systems. Concrete floors and walls can store heat in passively heated homes or livestock buildings. For storing heat in rocks, select a rock size between ¾" and 2". Rocks smaller than ¾" cause too much resistance to airflow, and rocks larger than 2" have a poor surface area/mass ratio—heat does not transfer into the rock centers very well. To minimize airflow resistance, use smooth rock of uniform size and construct the bed with a large face area and relatively short length in the airflow direction. Use Table 9 to estimate pressure drops through rock beds. Remember that most agricultural ventilation fans are designed to operate at a static pressure of ⅛" (0.125") water, or less. Solar systems usually require fans that operate at higher pressures.

Table 9. Pressure drop in rock beds.
Face velocity is airflow divided by face area.
Underlined numbers refer to Example 5.

Face velocity fpm	Rock diameter, in.			
	¾	1	1½	2
	Pressure drop			
	- - - - -in. of water/ft of bed length- - - - -			
10	0.01	0.0065	0.004	0.0025
15	0.02	0.012	0.007	0.005
20	0.03	0.02	**0.01**	0.0075
30	0.055	0.04	0.02	0.015
40	0.09	0.06	**0.035**	0.025
50	0.13	0.09	0.05	0.04
60	0.2	0.13	0.08	0.06
70	0.25	0.17	0.1	0.08
80	0.3	0.22	0.15	0.1
90	0.4	0.28	0.17	0.13
100	0.45	0.33	0.2	0.15

Values were interpolated from graph by Blaine Parker, University of Kentucky agricultural engineer.

Example 4:

How much energy is regained from a 100 ft^3 bed of 1½" rock cooled from 120 F to 50 F?

Answer:

From Table 8, rock has a volumetric specific heat, SHv, of 20 Btu/ft^3-F.

Qs = VOL x SHv x TD (Eq 3)
= 100 x 20 x (120 - 50)
= **140,000 Btu**
WT = 100 ft^3 x 100 lb/ft^3 = 10,000 lb

Note that rock stores less heat per cubic ft and per lb than water. Potential problems with rock beds include dust accumulations, water condensation in the bed, mold, odors, and algal growth on the rocks.

The storage container is usually made of lumber or concrete. It must be airtight and able to support the weight and sidewall pressure of the rock bed. Insulate rock beds to at least R=11 on all sides.

Example 5:

Estimate the pressure drop caused by 1000 cfm of air moving through a 5'x5'x10' bed of 1½" rock.

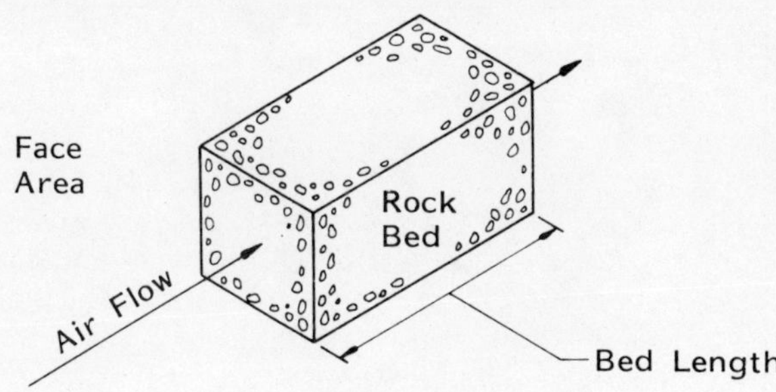

$$\text{Face Velocity} = \frac{\text{Air Flow}}{\text{Face Area}}$$

Answer:

Assume air enters the 5'x5' end.

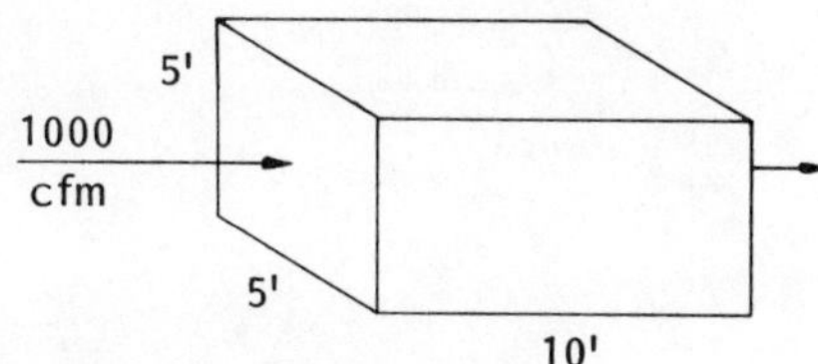

Volume = 5'x5'x10' = 250 ft³
Face velocity = 1000 cfm ÷ 25 ft² = 40 fpm
From Table 9, the pressure drop is 0.035" of water per ft of bed length.
Pressure drop = 10' x 0.035"/ft = **0.35"** of water.

Assume air enters the 5'x10' top.

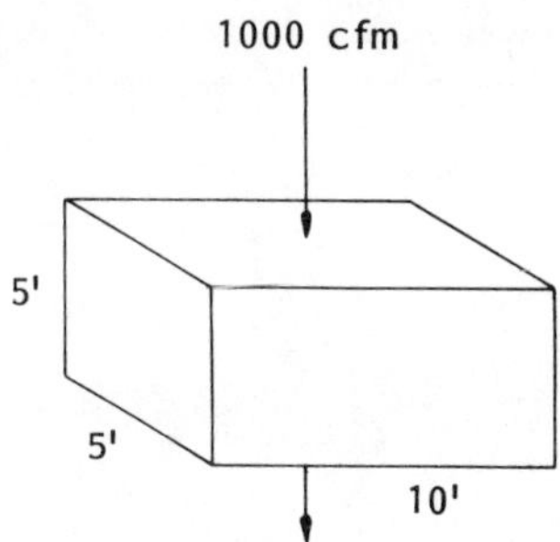

Volume = 5'x5'x10' = 250 ft³
Face velocity = 1000 cfm ÷ 50 ft² = 20 fpm
From Table 9, the pressure drop is 0.01" of water per ft of bed length.
Pressure drop = 5' x 0.01"/ft = **0.05"** of water.
Note that the pressure drop is 7 times as high with 10' instead of the 5' air path.

As you can see, rock bed shape and airflow direction have a large effect on pressure drop. Design your rock storage bed with reasonable air velocity and air travel distance through the bed.

Latent Heat Storage

When a solid is heated (sensible heat), its temperature rises until it reaches the melting point. At the melting point, any added energy (latent heat of fusion) changes the material from the solid phase to the liquid phase without raising its temperature. The same quantity of heat is released when the material freezes from a liquid back into a solid. Phase change salts have melting points and heats of fusion that make them useful for storing energy in solar heating systems. One of these, Glauber's salt ($Na_2SO_4 \cdot 10H_2O$, sodium sulfate decahydrate), melts at 91 F and has a heat of fusion of 108 Btu/lb. Its other properties are listed in Table 8.

The amount of heat that can be stored in these salts can be calculated from the following equations:

Eq 4.

To warm a solid:
Qs = WT x SH x TD

Eq 5.

To melt a solid:
QL = WT x LH

Eq 6.

To warm a liquid:
Qs = WT x SHL x TD

Qs = quantity of sensible heat, Btu
QL = quantity of latent heat, Btu/lb
WT = weight of storage material, lb
SH = specific heat of storage material, Btu/lb-F
SHL = liquid specific heat, Btu/lb-F
LH = latent heat of melting, Btu/lb
TD = temperature difference, F

Stacks of small, plastic containers or rows of tubes of phase change salt make up a latent heat storage bed. The heat transfer fluid passes through spaces between the containers.

Example 6:

How much energy is stored when 3500 lb of Glauber's salt is heated from 50 F to 120 F? How much space does the salt bed occupy?

Answer:

Properties of Glauber's salt are listed in Table 8. At 50 F, Glauber's salt is a solid. It stores sensible heat as it is warmed to 91 F.

Qs = WT x SH x TD (Eq 4)
   = 3500 x 0.5 x (91 - 50)
   = 71,750 Btu

At the melting point, the salt absorbs 108 Btu/lb without an increase in temperature until all the solid is melted.

QL = WT x LH
   = 3500 x 108
   = 378,000 Btu

Once all the solid is melted, the liquid stores additional sensible heat as an increase in temperature from 91 to 120 F.

Qs = WT x SHL x TD
   = 3,500 x 0.8 x (120 - 91)
   = 81,200 Btu

Total heat energy = 71,750 + 378,000 + 81,200
= **530,950 Btu.**

Note that most of the heat stored is latent heat.

A salt bed occupies about twice the volume of the liquid salt, allowing for space between the salt containers.

$$VOL = 2 \times WT \div DEN$$
$$= 2 \times 3500 \div 70$$
$$= 100 \text{ ft}^3$$

VOL = volume of storage material, ft³
WT = weight of storage material, lb
DEN = density of storage material, lb/ft³

More energy can be stored in a cubic ft of phase change salts than in rock or water. See examples and Table 8. Much heat energy can be removed from or added to storage without changing the storage temperature. With sensible heat storage, the storage temperature and heating system performance drop as heat is removed. The advantages of phase change salts are offset by their high cost and uncertain life. **Some salts tend to break down and do not completely freeze after a number of freeze-thaw cycles.**

Utilizing Heat Storage

Many well-insulated, fully occupied livestock buildings need little supplemental heat during daytime hours when the sun is shining and temperatures are at their maximum. They frequently need heat at night when outside temperatures reach their daily minimum. Heat storage allows solar energy use when it is needed most and increases the collector system's efficiency.

A solar system designed to meet 100% of a building's heating load in all weather conditions (including extreme cold at the end of an extended cloudy period), would be very large and expensive and have excess solar capacity most of the time. It is generally more cost-effective to design a solar system with storage which provides enough heat for about 36 hr. For heating needs beyond that, provide a backup heater. Although the heater runs less often when storage is included in the solar heating system, the heater must be sized for the maximum calculated heat loss. On cold, cloudy days after stored heat is depleted, the backup heater must handle the full heating load.

In solar energy systems that preheat ventilating air for a livestock building, one of the important functions of the heat storage is to produce a time lag in the delivery of solar heat. During the day, outdoor air plus solar energy warm the storage material. Later, the cool night air removes heat from storage to help maintain the building temperature. The heat storage unit has two effects:

1. It smooths out the wide fluctuations in solar-heated air temperatures, Fig 17. A solar collector without storage increases natural day-night temperature fluctuations.
2. It increases the number of days that solar energy can heat the building. In winter, air leaving a solar collector may be below the desired building temperature, even on sunny days. But on sunny spring and fall days, a solar collector can overheat the building, so

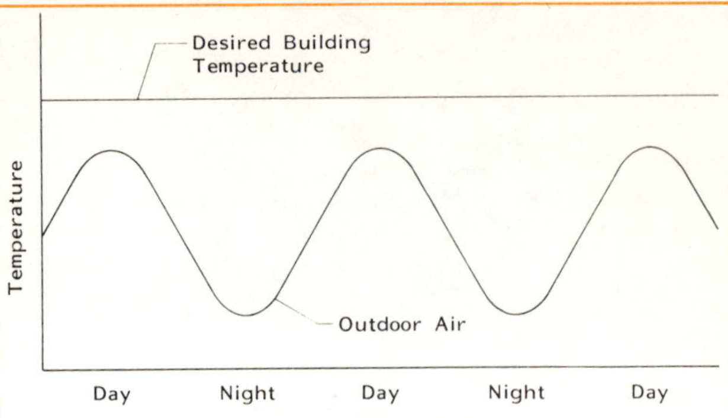

17a. Typical winter day-night outdoor temperatures.

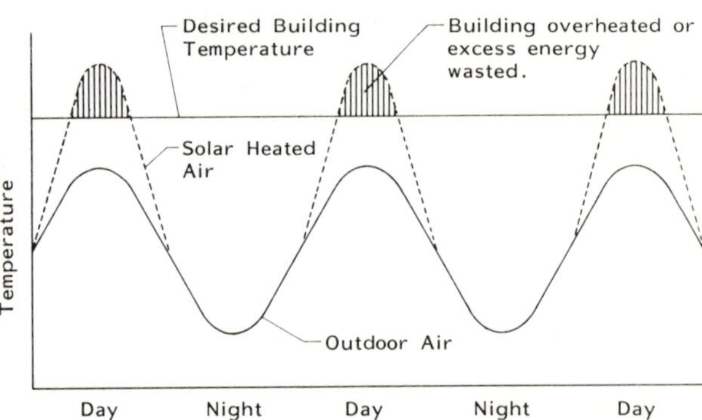

17b. Collector output temperatures without storage.

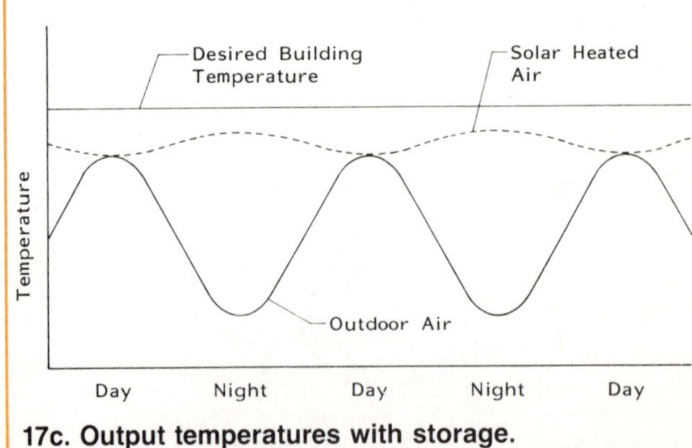

17c. Output temperatures with storage.

Fig 17. Temperature of solar preheated ventilating air.

the collector must be bypassed. With properly sized heat storage, however, the excess daytime heat is stored until evening, and the solar system can operate all day without overheating the building. A greater total amount of energy is utilized, more fuel is saved, and solar system efficiency is increased with storage.

Solar Systems

There are two basic types of solar systems—passive and active.

Passive Systems

In passive solar systems, heat is generally transferred without pump or fan power. Natural convection, conduction, and radiation distribute the collected solar energy.

The Trombe wall is one example of a passive solar heating system, Fig 18. The heart of the system is a massive, energy-storing concrete or masonry wall on the south side of the building. The outside of the wall is painted black and has a transparent cover. Solar radiation warms the wall during the day. Natural convection moves heated air past the front of the wall and into the building. Heat is also radiated into the building from the inside wall surface. Variations are possible—columns of water can replace the solid wall, or in combination passive and active systems, fans rather than natural convection can move the air past the wall.

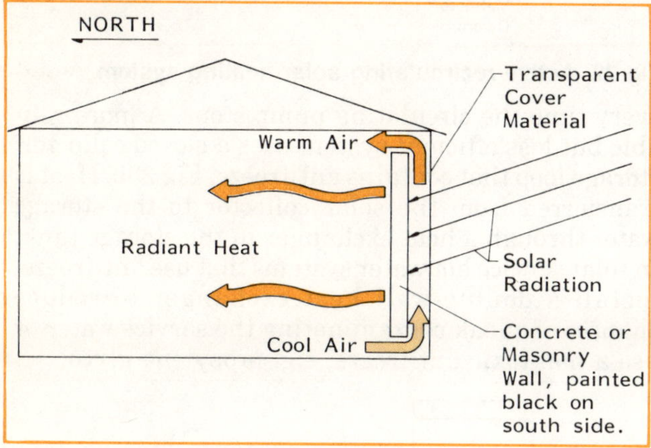

Fig 18. Passive solar heating with a Trombe wall.

Direct gain solar heating is also passive, Fig 19. Solar energy entering the building through south windows is absorbed by the materials inside. At night, these materials help keep the building warm. Heating is more effective if the back wall and floor of the structure are concrete (or other good heat storage materials). Window area is based on the heating needs for an average winter day. With direct gain heating, insulate the south windows at night or you can lose at least as much heat as you gained during the day. South facing, open-front livestock buildings utilize simple direct gain solar heating with a roof overhang that allows the winter sun to penetrate and keeps the summer sun out, Fig 20.

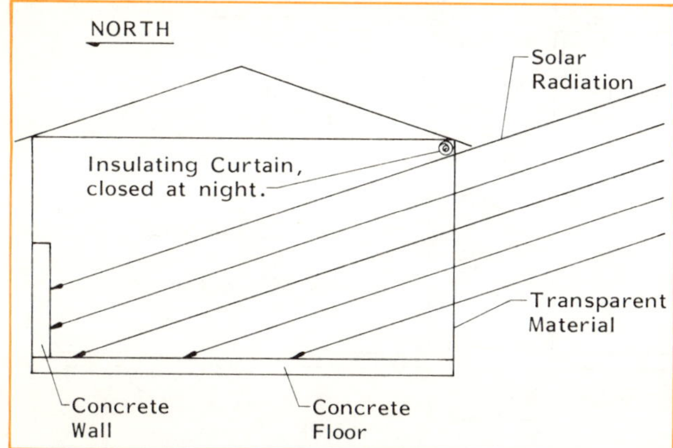

Fig 19. Passive solar heating by direct gain.

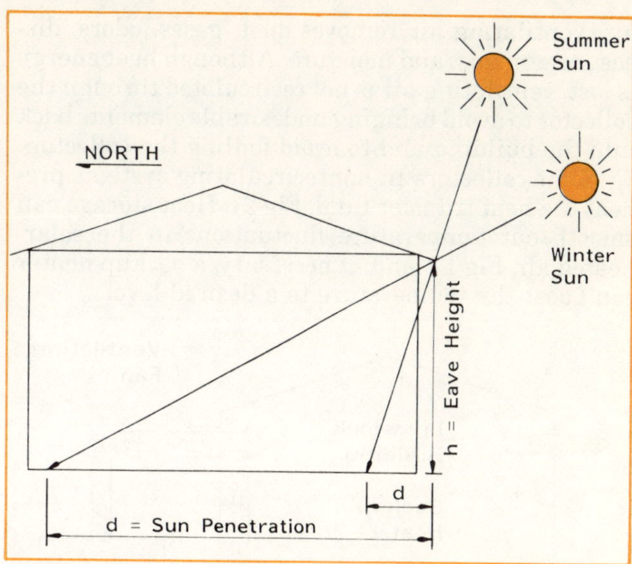

Fig 20. Direct solar gain in an open-front building.

You can estimate sun penetration with the solar angle factors in Table 1. Although sunlight can go deep into a building in early morning and late afternoon, little significant heating occurs.

Example 7:

Find the maximum sun penetration into a south-facing, open-front building with a 10' eave height between 9 a.m. and 3 p.m. on both June 21 and Dec. 21 at 40° north latitude.

Answer:

Maximum 9 a.m. to 3 p.m. penetration occurs at 9 a.m. and 3 p.m. on Dec. 21 and at solar noon on June 21. Use the solar angle factors, SAF, from Table 1. SAF for June 21 is 0.3 and SAF for Dec. 21 is 3.0.

$Pen = SAF \times h$

Pen = sun penetration, ft
SAF = solar angle factor
h = eave height, ft

Summer penetration
= 0.3 x 10'
= **3.0'** (at solar noon)

Winter penetration
= 3.0 x 10'
= **30.0'** (at 9 a.m. and 3 p.m.)

Passive solar systems can be simple and can provide heat at low operating and maintenance costs. Disadvantages include poor temperature control with wide temperature fluctuations in the heated space.

Active Systems

In active solar systems, pumps or fans move the heat transfer fluid. Some applications require extra pumps or fans, but others such as preheating livestock building ventilating air, use existing fans. The airflow resistance in the solar collector slightly reduces system airflow.

The heat transfer fluid in solar systems on livestock buildings is usually fresh, or non-recirculated, air. Ventilating air removes dust, gases, odors, disease organisms, and moisture. Although heat energy is lost, ventilating air is not recirculated through the collector to avoid bringing undesirable elements back into the building and to avoid fouling the collector.

Solar collectors in nonrecirculating systems preheat the heat transfer fluid, Fig 21. Heat storage can smooth out temperature fluctuations in the solar-heated air, Fig 17, and, if necessary, a backup heater can boost the temperature to a desired level.

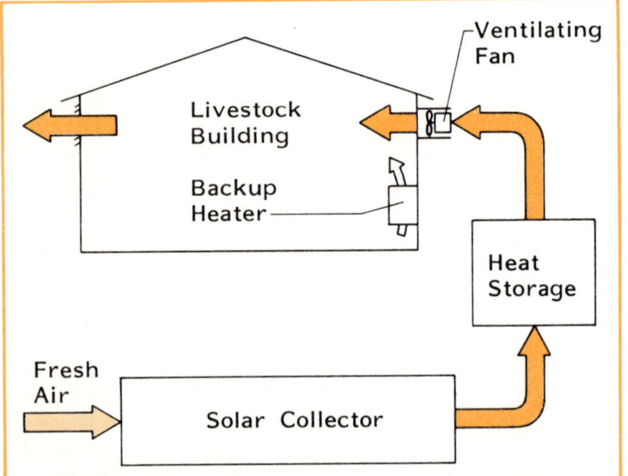

Fig 21. Solar preheated nonrecirculating system.

Solar air preheaters can be quite simple and still operate at high efficiency, because the difference between the heat transfer fluid temperature and the outdoor temperature is usually small. Collector efficiency is highest if all the ventilating air passes through the collector. If only part of the air is solar-heated:

$$TD = Scfm \times STD \div Tcfm$$

TD = temperature rise of total system airflow, F

Scfm = solar heated airflow (cfm)

STD = solar temperature rise, F

Tcfm = total system airflow (cfm)

Some active systems recirculate the same air or liquid through the solar collector. Recirculating systems, Fig 22, are used mainly to heat shops and homes and can be used for either service water or heating water. Fluid entering the collector is usually above room air temperature. Output temperatures can be high, but only high quality collectors are efficient. Seal air recirculating systems tightly to prevent cold air infiltration and efficiency loss. Also, avoid running the collector if the inlet fluid's temperature is higher than the absorber's, or heat will be lost from the system. Install a differential thermostat to sense fluid and absorber temperatures and switch on the pump or fan in the collector loop only when the absorber temperature is high enough.

Water is the usual storage material for liquid-type solar systems. In drain-down systems, water from storage is circulated through the collector, Fig 23a. To keep the water from freezing and rupturing the collector tubes, the collector is automatically drained

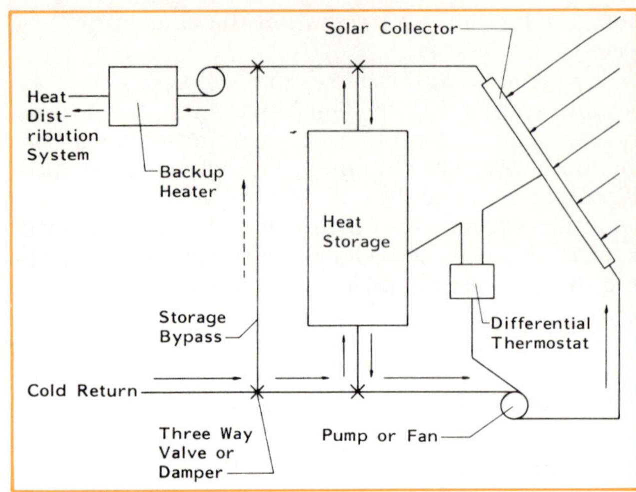

Fig 22. Active recirculating solar heating system.

every time the circulating pump stops. A more reliable but less efficient system uses a closed collector/storage loop that contains antifreeze, Fig 23b. Heat is transferred from the solar collector to the storage water through a heat exchanger in the storage tank. In solar service hot water systems that use antifreeze, install a double-wall heat exchanger to reduce chances of a leak contaminating the service water or use a non-toxic antifreeze, e.g. propylene glycol.

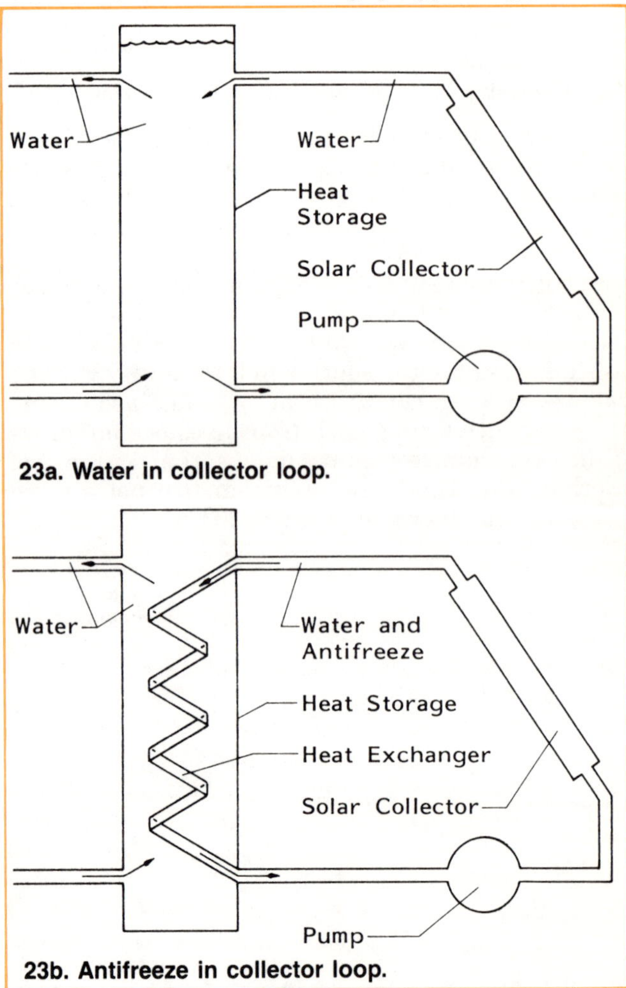

23a. Water in collector loop.

23b. Antifreeze in collector loop.

Fig 23. Liquid solar systems.

On sunny days, when the collector is not in use and the liquid is stagnant (not moving), temperatures inside the collector can exceed 300 F. Protect against damage from boiling liquid by draining the collector or providing a pressure relief valve. In addition to boiling and freezing problems, liquid-type systems are prone to corrosion, scale buildup, fluid leakage, and algal growth.

Flat Plate, Air-Type Collector Construction

There are many variations of air-type solar collectors for new or retrofit building construction and for freestanding and portable units. Some examples are shown in the plans in this handbook. These guidelines will help you build efficient, long-lasting collectors:

- Select materials that withstand ultraviolet radiation, heat, wind, snow, rain, and hail; are easy to install; and have long life and low cost. Many common building materials (sheet metal, chipboard, plywood, dimension lumber) are adequate.
- In collectors with distinct air channels, keep collector air velocity between 500 fpm and 1000 fpm. Air velocity (fpm) equals airflow through the collector (cfm) divided by the cross sectional area of the collector (ft²). At air velocities under 500 fpm, simple collectors may be inefficient; above 1000 fpm, pressure drop is usually too high, and fans require too much power.

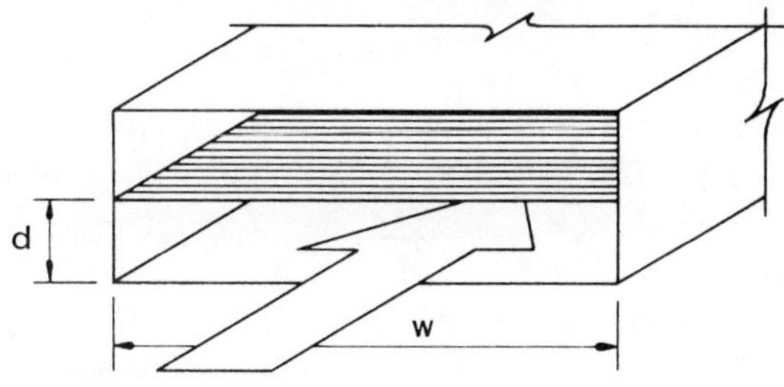

Cross-sectional Area = d x w

$$\text{Air Velocity (fpm)} = \frac{\text{Air Flow (cfm)}}{\text{Cross-sectional Area (sq ft)}}$$

- To prevent excessive pressure drop in ducts, keep ducts short and straight and keep air velocity below 800 fpm. Size ducts for about 800 cfm/ft² of duct area.
- Keep static pressure drop in collector and ducts less than 1/8″. Make sure that your fan will deliver adequate air in solar systems with higher air resistance, such as with heat storage in rock. Greater pressure drops reduce fan performance too much. If pressure drop is excessive, bypass the solar system with part of the air, rebuild the collector and/or ducts, or install centrifugal fans designed for higher pressure. (Pressure drop is the difference between the inlet and outlet air pressure. In nonrecirculating systems where the fan draws outdoor air through the collector, the inlet pressure is 0—just measure the negative pressure at the entrance to the fan.)
- Cover air inlets with ½″ hardware cloth to keep out birds, rodents, and trash.
- Use pressure preservative treated lumber for components in contact with the soil. Use CCA treatments to avoid the volatiles in oil borne preservatives.
- Seal the collector to prevent air leaks. Caulk joints and cracks with silicone or butyl rubber and seal cover-sheet laps with clear silicone. Other types of caulking are cheaper but do not last as long or work as well on solar collectors. Seal the edges of corrugated materials with corrugated wood or closed-cell foam strips.
- Fasten FRP with screws rather than nails.
- Position rows of portable or freestanding collectors so they do not shade one another. Maintain adequate distance between buildings to prevent shading.
- If collectors are not used during the summer, shade them from the sun and/or open vents (1 ft² of vent opening per 100 ft² of collector surface area) to keep them cool. Locate vent openings to cool the collector with prevailing winds or natural convection.

Basic Equations for Collectors

Use the following equations to size solar collectors or predict collector performance. (Note: collector area equals gross collector area based on outside dimensions.)

Definitions follow for Eqs 7 through 11:

$$
\begin{aligned}
1.1 &= \text{a constant to allow for air density and} \\
    &\quad \text{specific heat of air, Btu/F-hr-cfm} \\
A &= \text{collector area, ft}^2 \\
\text{Ccfm} &= \text{collector airflow, cfm} \\
\text{CE} &= \text{collected energy, Btu/hr} \\
\text{EFF} &= \text{collector efficiency, \% (For efficiency as} \\
    &\quad \text{a decimal, do not multiply by 100.)} \\
\text{RAD} &= \text{solar radiation, Btu/hr-ft}^2 \\
\text{TD} &= \text{temperature rise, F}
\end{aligned}
$$

Eq 7.

Temperature rise through a collector:
TD = EFF x RAD x A ÷ (1.1 x Ccfm)

Eq 8.

Collector airflow to get temperature rise:
Ccfm = EFF x RAD x A ÷ (1.1 x TD)

Eq 9.

Collector efficiency:
EFF = 1.1 x Ccfm x TD ÷ (A x RAD)

Eq 10.

Collector energy output, by airflow:
CE = 1.1 x Ccfm x TD

Collector energy output, by efficiency:

Eq 11.

$$CE = EFF \times A \times RAD$$

The temperature rise used in the equations equals outlet temperature minus inlet temperature and can be either maximum or average. Maximum or peak temperature rise, for collectors without integral storage (no rock or concrete in the collector), occurs near solar noon on sunny days. Use noon hour, clear day solar insolation from Table 26, appendix, and noon hour efficiency when working with peak temperature rise. Measure peak temperature rise with the fan or pump running and thermometers shaded.

Twenty-four hour average temperature rise cannot be measured directly. It is the temperature rise that would occur if daily solar radiation arrived at a constant rate over 24 hours. When working with average temperature rises, use average daily insolation from Table 27, appendix, and average collector efficiencies.

Example 8:

Find the collector airflow required to produce 110 F output temperatures from a 96 ft² solar collector at noon on a sunny Dec. 21 at Lemont, IL. Collector inlet temperature is 65 F; the tilt angle is 60°; and noon hour efficiency is 45%.

Answer:

Lemont is at 41.4° north latitude, so use data for 42°. A 60° tilt angle is at about latitude plus 15° at 42° north latitude. From Table 26, appendix, solar insolation equals 287 Btu/hr-ft² at noon.

$$TD = 110\ F - 65\ F = 45\ F$$

Using Eq 8,

$$Ccfm = 0.45 \times 287 \times 96 \div (1.1 \times 45)$$
$$= \textbf{250.5 cfm}$$

Select a fan to deliver 250 cfm at about ¼″ pressure.

Example 9:

Find the noon hour rate of energy output for the collector in Example 8.

Answer:

Solution 1: Calculate energy output from Eq 10:

$$CE = 1.1 \times 250.5 \times 45\ F$$
$$= \textbf{12,400 Btu/hr}$$

Solution 2: Knowing solar input and collector efficiency, calculate collector output from Eq 11:

$$CE = 0.45 \times 96 \times 287$$
$$= \textbf{12,398 Btu/hr}$$

Example 10:

Find the expected 24-hr average December temperature rise for the collector in Example 9. Average collector efficiency is 40%; collector airflow is 250 cfm.

Answer:

From Table 27, appendix, the average December insolation on a 60° surface (about latitude plus 15°) in Lemont is 964 Btu/day-ft².

$$RAD = 964 \div 24\ hr/day$$
$$= 40.2\ Btu/hr\text{-}ft^2\ (average)$$

Using Eq 7,

$$TD = 0.4 \times 40.2 \times 96 \div (1.1 \times 250)$$
$$= \textbf{5.61 F}$$

Example 11:

Find the expected average December daily energy output for the collector in Example 10.

Answer:

Solution 1 using Eq 10:

$$CE = 1.1 \times 250 \times 5.61 \times 24\ hr/day$$
$$= \textbf{37,000 Btu/day}$$

Solution 2 using Eq 11:

$$CE = 0.4 \times 96 \times 964$$
$$= \textbf{37,000 Btu/day}$$

Example 12:

What is the December daily fuel oil equivalent of the energy produced by the collector in Example 11?

Answer:

From Table 20, we see fuel oil contains 138,000 Btu/gal and is typically burned at 65% efficiency.

$$Oil = CE \div (Qoil \times Beff)$$
$$= 37,000 \div (138,000 \times 0.65)$$
$$= \textbf{0.41 gal/day}\ \text{in December}$$

Oil = fuel oil equivalent, gal/day
Qoil = energy in fuel oil, Btu/gal
Beff = burner efficiency, decimal

SYSTEM FLOW DIAGRAMS

This section shows some of the options for collecting, storing, and distributing solar energy for farm buildings. The schematic diagrams show the major components and the movement of air or liquid to transport the heat. The examples illustrate some of the ways the components are put into buildings.

Preheating Ventilating Air

Relatively simple and inexpensive solar systems can preheat ventilating air. For livestock buildings, do not recirculate room air. Liquid-type collectors are avoided because they require heat exchangers and are relatively expensive.

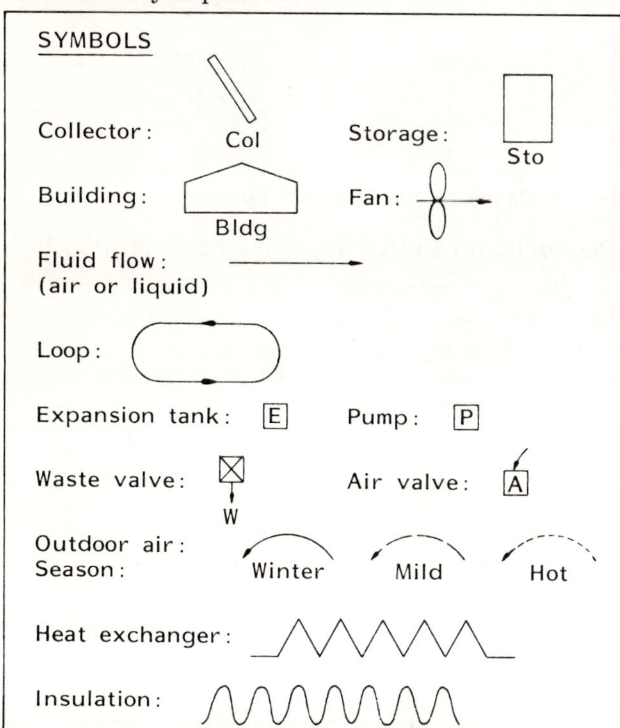

One Pass System without Storage (Fig 24)

In winter, air passes through the collector, where it is warmed, and goes directly to the heated space. In mild weather, some fresh air is blended into the warmed air to prevent overheating of the building. In summer, bypass the collector, but protect it from overheating by covering or venting it.

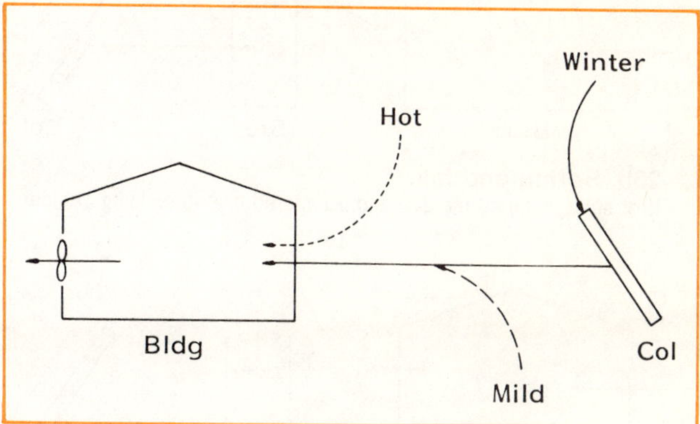

Fig 24. One pass system without storage.

This simple system has limited value, because it can overheat the building in warm weather unless the collector is undersized. See Fig 17 in the Solar Energy Storage section, Fundamentals. It is useful for retrofitting buildings with little insulation and in other cases where daytime heat is needed. (Adding insulation can be more cost-effective.) Applications include some swine growing buildings, colder climates, and retrofitting poorly insulated buildings for smaller animals. The system is also applicable to nursery and brooder buildings requiring summer heat.

If solar heating is added to a building that could be naturally ventilated, the fan energy needed to move the solar heat can exceed the useful energy gained.

Examples of one pass systems without storage include a solar attic without storage, a covered plate collector on the south wall, and some installations with portable or commercial collectors.

One Pass System with Storage (Fig 25)

Storage increases use of solar heat, levels out temperature cycles in the warmed space, and permits some summer cooling. These systems permit using less solar heat in the daytime and saving that heat for nighttime, when it is more typically needed. Well insulated buildings are desirable.

For summer cooling, pass night air through the collector and into the storage to cool it and exhaust it outdoors, Fig 25c. During the day, and until the storage warms up, bring ventilation air through the storage to the building space, Fig 25d.

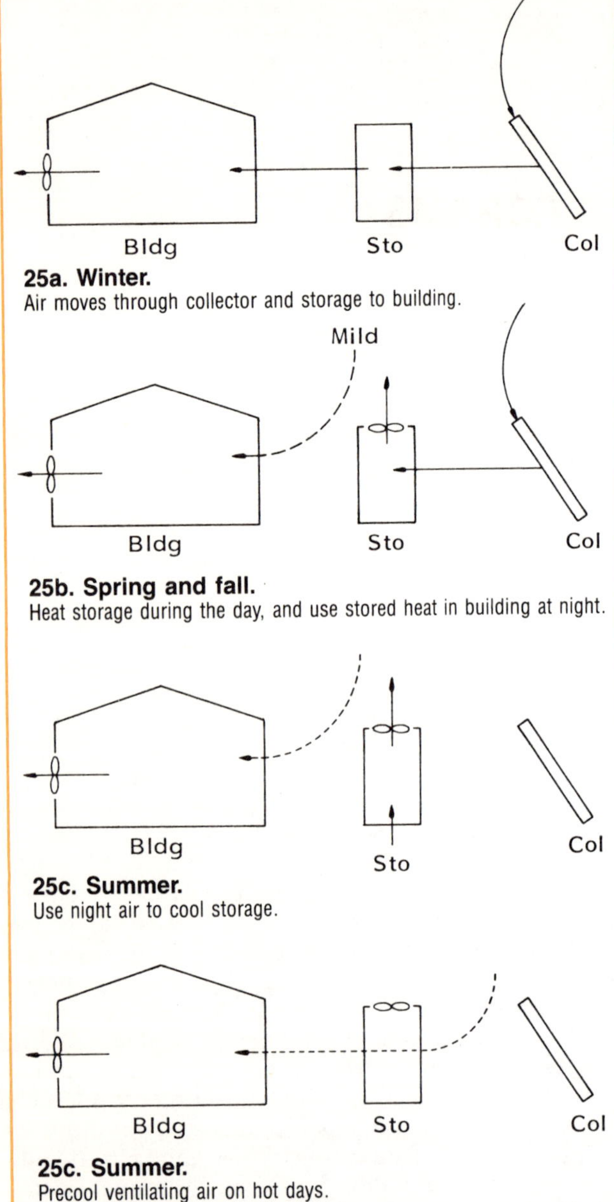

25a. Winter.
Air moves through collector and storage to building.

25b. Spring and fall.
Heat storage during the day, and use stored heat in building at night.

25c. Summer.
Use night air to cool storage.

25c. Summer.
Precool ventilating air on hot days.

Fig 25. One pass system with storage.

Collect/Store Loop, One Storage/Bldg Pass (Fig 26)

In this system, the temperature of the storage is raised by recirculating air between the collector and the storage during daylight hours. In mild weather, heat can be stored while the ventilating air is bypassing the storage, permitting more heat collection during mild weather in anticipation of cold nights.

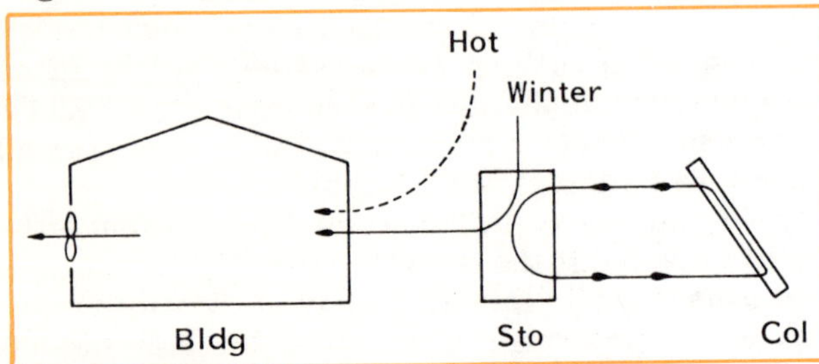

Fig 26. Collect/store loop, one storage/building pass.

Space Heating

Space heating is supplied by solar energy that is not added to the ventilating air.

Higher temperature collection and storage permits space heating of homes, shops, offices, milking centers, etc. These systems are suitable for buildings that do not require the high ventilating rates of typical livestock housing and that have relatively clean air that can be recirculated through storage. High temperature recirculating collectors are also useful for zone heating portions of livestock buildings.

Passive solar heating of livestock housing is another example of solar space heating.

Passive Solar Livestock Buildings (Fig 27)

Part of the south wall and/or roof allows sunlight to warm and dry the building directly. This is the simplest solar system, because no separate equipment or controls are needed. Some heat is stored in the building materials, especially in concrete walls and floors.

Some management is needed to conserve captured energy and to prevent seasonal overheating. Cover the solar glazing at night to reduce heat loss through it. Night losses can exceed the daytime gains unless the glazing is insulated at night. Shade the glazing in summer to prevent overheating.

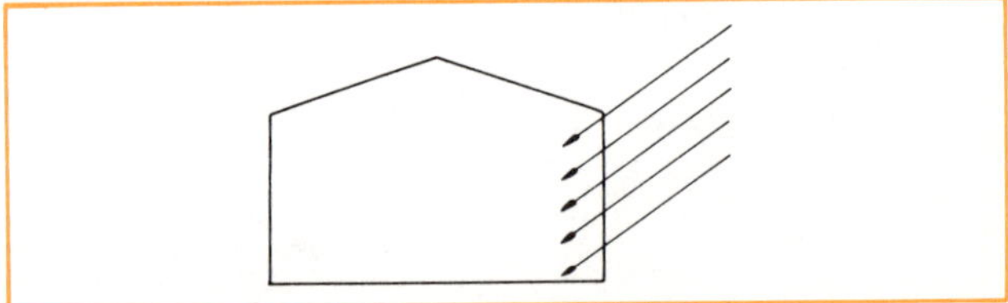

Fig 27. Passive solar livestock buildings.

Recirculating Air Heating without Storage (Fig 28)

This is a relatively simple system suitable for heating homes or shops but not livestock housing. It collects heat only during sunlight hours. Because there is no storage, it does not contribute heat at night or during cloudy days. If the collector is too large for the space being heated, overheating will result.

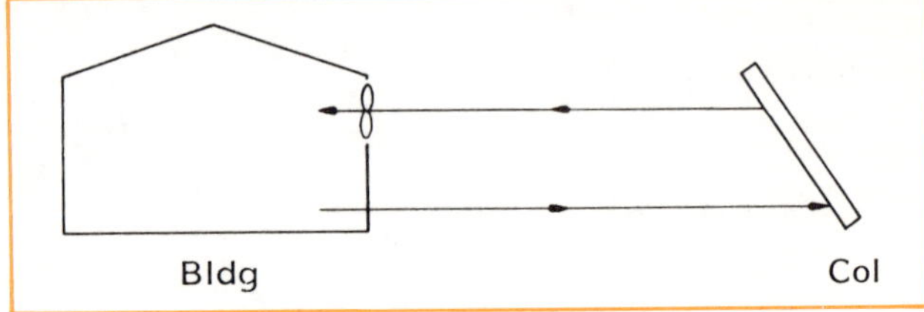

Fig 28. Recirculating air heating without storage.

Indirect Space Heating from Storage (Fig 29)

Heat from the collector can be stored in a rock bed or other storage beneath a floor. The room is heated directly by radiation, or resting animals are heated by conduction from the warm floor surface.

Circulate air through the collector and storage only when the temperature of the air coming out of the collector is higher than the temperature in the storage. Turn off the collector when storage reaches maximum temperature, or when the room is warm enough.

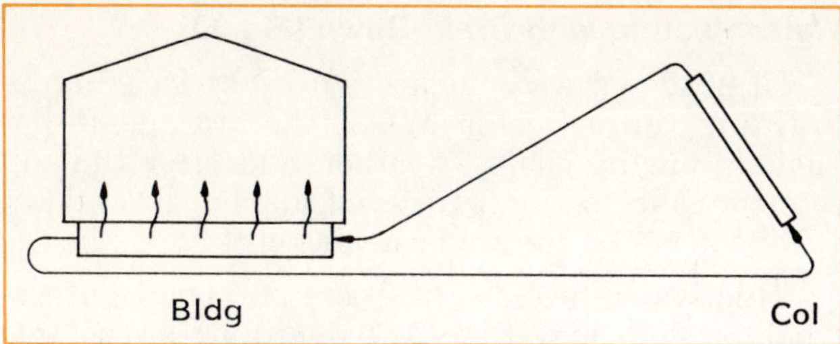

Fig 29. Indirect heating from storage.

Recirculating air system with storage & bypass (Fig 30)

In this system, the building is heated by circulating room air through the storage or the collector.

Circulate air through (see Fig 30):
a. Collector and storage when the sun is shining and the building needs no heat.
b. Building and storage when the building needs heat but the sun is not shining.
c. Collector and building when the building needs heat and the sun is shining.
   Or,
d. Fig 30d shows a preferred system that circulates air through both the collector and storage when the sun is shining and the building needs heat. This mode requires more complicated ducts and dampers. Drawing heat from storage provides more uniform ventilating air temperature.

Usually air from the collector goes through the storage in one direction, while air for the building goes through in the other direction. More uniform temperatures in the ventilating air result.

Liquid-Type Collectors for Space Heating

One loop carries heat in the pumped fluid from the collector to the storage. Another loop carries heat from the storage to the building.

Heat from storage can be distributed through the building space with fin tubes, pipes in floor, fan and radiator, heat exchanger in the ventilating duct, or other liquid/air heat exchangers. This system is useful for residential, greenhouse, or milkhouse air heating and for some swine buildings.

Insulate the top, bottom, and all sides of the storage tank to at least $R = 11$. Provide good drainage around buried storage tanks to prevent flotation and to reduce heat loss. Also provide an expansion tank in any closed system to allow for volume changes in the liquid caused by temperature changes.

Liquid system with drain-down (Fig 31)

A collector may freeze when there is no solar energy. One solution to the problem of freezing is to drain the collector when it starts to cool down.

In an open-loop system (Fig 31a), one pump delivers hot water to the building and a separate pump

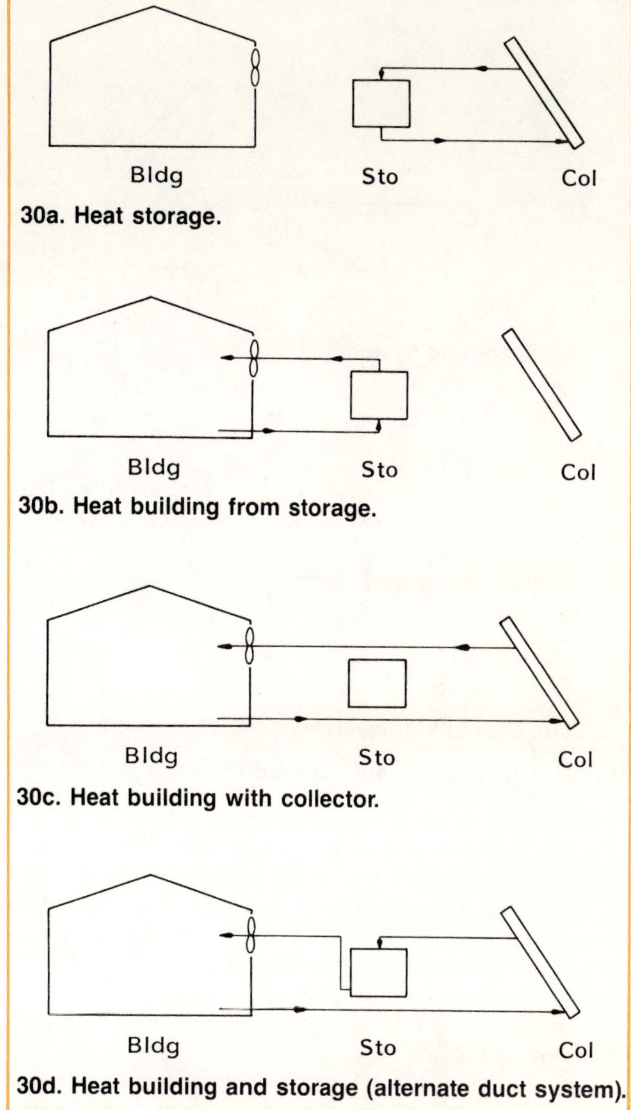

30a. Heat storage.

30b. Heat building from storage.

30c. Heat building with collector.

30d. Heat building and storage (alternate duct system).

Fig 30. Recirculating air system with storage and bypass.
Not suitable for livestock housing.

circulates fluid through the collector loop. There is an air gap between the surface of the liquid in the heat storage tank and the end of the collector loop return line. When the collector pump turns off, the air inlet valve opens, and the collector loop drains into the heat storage tank.

Install an air inlet valve at the high point of the collector loop so the whole loop drains. Also, locate the valve where it will not freeze shut—for example, inside the collector case. Because air enters the system, collector tube corrosion could be a problem.

In a closed-loop system (Fig 31b), one pump distributes hot water from heat storage and another circulates water through the collector loop. Install a drain-down valve in the lines to and from the collector. When the pump shuts off, the drain-down valve isolates the collector loop from the rest of the system. The air inlet valve opens, and the collector loop drains to waste.

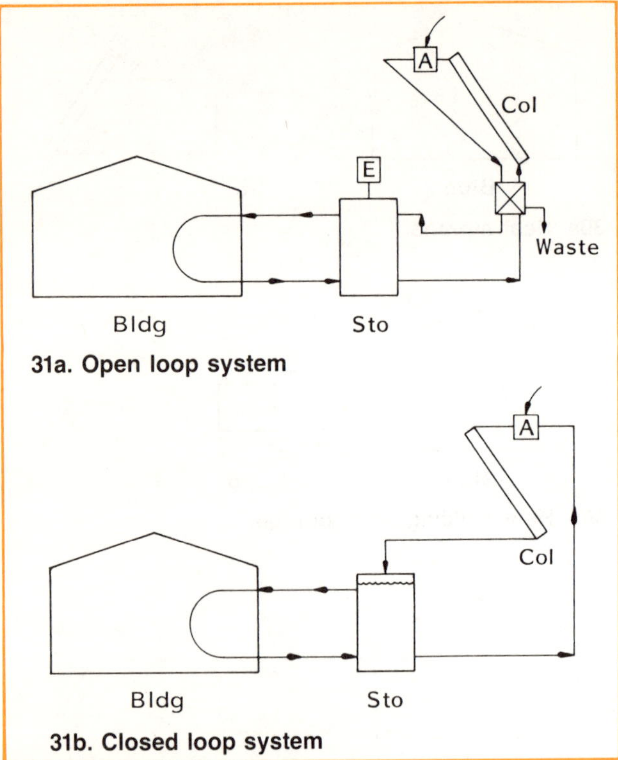

31a. Open loop system

31b. Closed loop system

Fig 31. Drain-down liquid system.

Liquid system, antifreeze in collector loop (Fig 32)

By adding a heat exchanger in the collector loop, it can have antifreeze while the rest of the system uses plain water. The heat exchanger must be guaranteed to protect against antifreeze leakage, the antifreeze must be nontoxic, or the liquid in the rest of the system cannot be used for drinking or other pure-water purposes. Expect lower efficiencies than for the system in Fig 31.

Some systems with antifreeze mix the collector-loop fluid and the room-loop fluid in the storage. Mixing avoids heat exchangers but requires that all the fluid be protected with antifreeze.

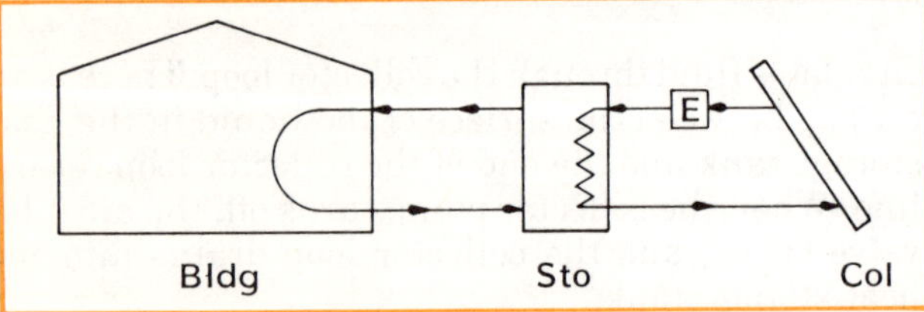

Fig 32. Liquid system with antifreeze in collector loop.

Service Water Heating

Solar energy can heat water for the home, milkhouse, swimming pool, etc. Packaged commercial systems are available.

Because solar water heating is not the most cost-effective way to use solar energy, consider other heat sources first, such as compressor heat from a milk cooler. Also consider other uses for the solar energy.

Water Heating with Drain-Down (Fig 33)

Drain-down water heaters prevent freezing by draining the collector. When the storage is hot enough, during cloudy weather, and after the sun sets, the collector pump turns off, and the drain-down valve opens and the water drains out.

This system avoids antifreeze problems, but the whole system may have to be suitable for potable water.

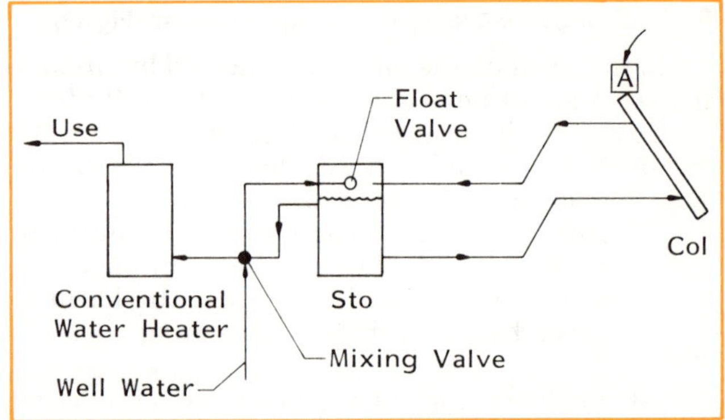

Fig 33. Service water heating with drain-down protection.
Hot water should be taken from top of tank, cold into bottom.

Water Heating with Antifreeze (Fig 34)

Heat from the collector warms water in the insulated preheater through a heat exchanger. The prewarmed water supplies a conventional water heater.

Commercial units are available which tend to be about 20% efficient. Use non-toxic antifreeze or a double-wall heat exchanger to assure no contamination of potable water if storage fluid is delivered directly to the water heater.

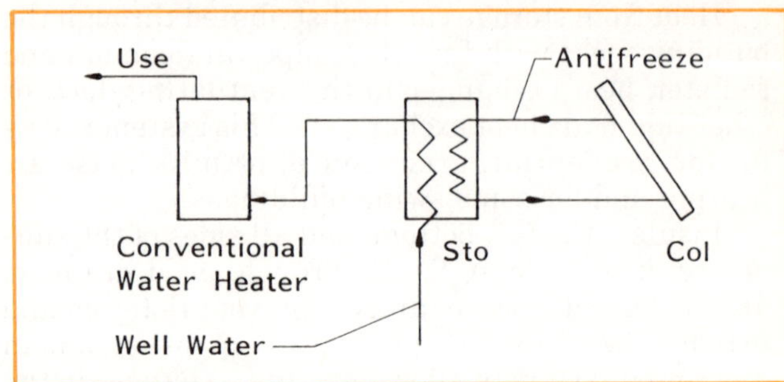

Fig 34. Service water heating, with antifreeze protection.

CONTROLS

The following is a summary of the types of controls that are useful in solar systems on farm buildings. Most of the components are commercially available. Look for quality, economy, cleanability, accuracy, and availability of adequate service.

Thermostats

Thermostats sense temperature and then open or close a switch to make something happen or stop happening. A **heating thermostat** closes a switch as the temperature falls, as for starting a furnace. A **cooling thermostat** closes a switch as the temperature rises, as for starting an air conditioner or ventilating fan.

Select a thermostat that matches the voltage of the system—12, 24, 120, or 240—and the amperage. Select one with a sensitivity range that matches needed accuracy—one or two degrees of error usually won't matter, but you will want to sense differences as little as 5 F.

Differential thermostats sense the difference between two temperatures. They are available with both fixed and adjustable differentials, but some are not very reliable. One application is to compare absorber plate temperature with storage temperature. If the absorber plate is at least 10 F above the actual storage temperature, the circulating pump turns on to move the working fluid through the collector which moves heat into storage. The thermostat turns off the pump when the temperature difference is down to about 6 F.

Thermometers

Thermometers in water and air lines and in storages help you learn how your system works, if it is working, and how it responds to various weather and building conditions.

Fittings are available for inserting a thermometer into a water pipeline. An indoor/outdoor thermometer can show the temperature within a storage. Remote sensing thermometers are available at electrical supply stores. Recorders are available that make a paper tape of temperature, or other readings, at several locations. They are expensive but can be helpful in a large installation.

Valves (Fig 35)

- **Flow control** valves balance the various parts of the system. They are manually set to limit flow below some maximum. Gate and ball valves contribute less to pressure loss in the system than do globe valves.
- **Solenoid** valves respond to a voltage. An electromagnet opens or closes a valve when energized by a circuit from a thermostat (or other control switch).

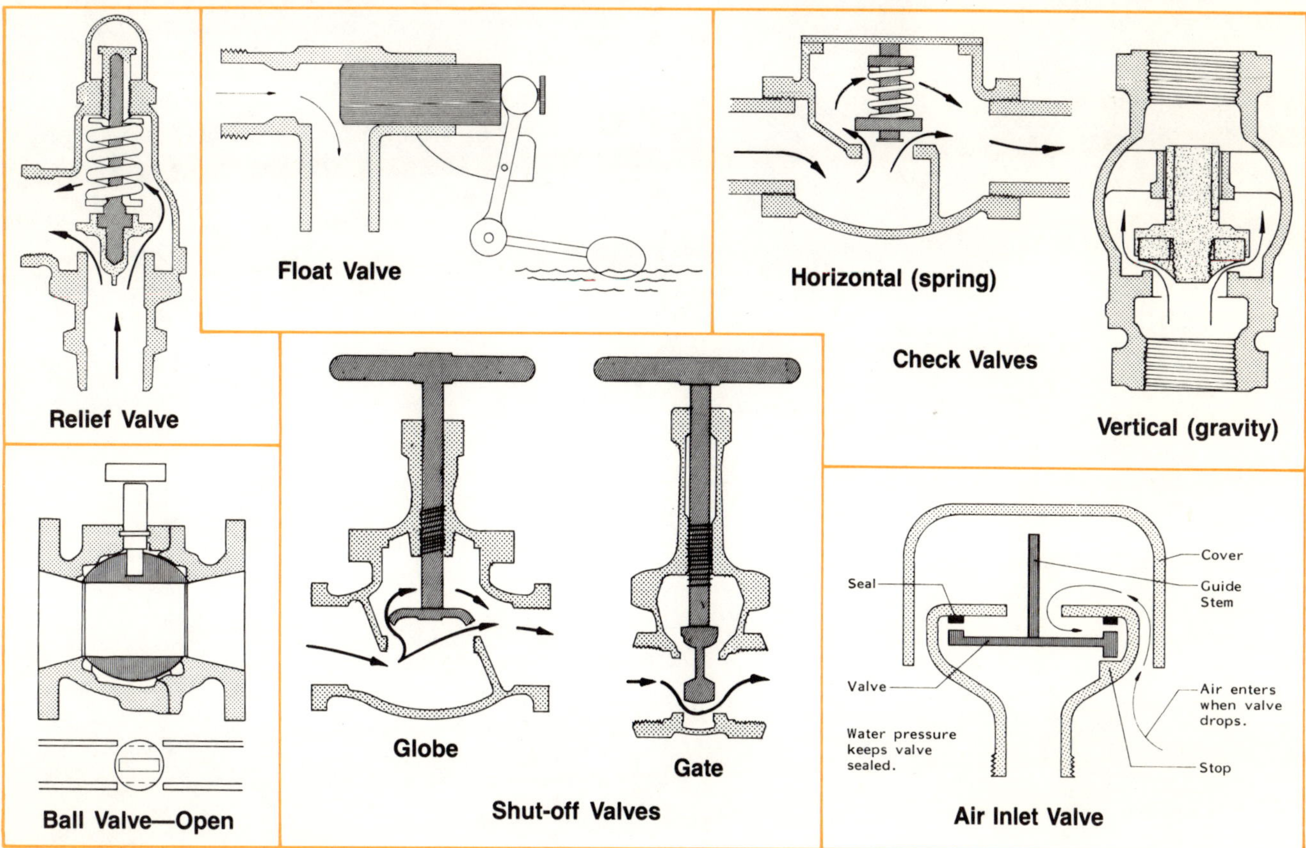

Fig 35. Typical valves.

- **Mixing valves** blend hot and cold liquids, like a typical domestic faucet. A type useful in solar installations responds to a thermostat to automatically blend cold water if a storage gets too hot (about 140 F).
- **Drain-down system valves.** Air inlet valves open when there is not water pressure against them. They let air into the high point of the collector loop so the lines can drain. See liquid system with drain-down, Fig 31. In an open-loop system, the air inlet valve opens when the pump turns off. In a closed-loop system, a mechanically operated drain-down valve opens when the circulation pump turns off. The air inlet valve also opens, and the collector loop drains to waste. Solenoid valves are recommended because they are more likely to work in bad weather.
- **Check valves** prevent liquid from flowing backwards in a line.
- **Pressure relief valves** permit fluid to escape from the system to relieve excess pressure. They are spring controlled and usually are adjustable. Install a relief valve or an expansion chamber in any closed hot water system.

Expansion Chamber

Install an expansion chamber in any closed-loop liquid system. It allows for the change in volume of the liquid as temperatures change.

Air Controls

- **Dampers** open or close a duct and can also control flow in partly-open positions. They can be motor driven in response to a thermostat or manually set.
- **Diverters** at a "Y" in a duct cause flow down one leg or the other of the Y.
- **Shutters** are louvered dampers. They can be power driven in response to a thermostat or other signal, or closed by gravity and opened by a moving airstream. Gravity shutters can serve as a check valve by reducing backward flow.

Alarms

Consider a high temperature alarm in a water storage, especially with a recirculating collector loop which can create high temperatures. A high temperature alarm in a collector warns you of needed summer ventilation to prevent overheating. A low temperature alarm warns you of heating system failures in housing for young animals or that the building or solar system may freeze.

A general power failure alarm is important for livestock installations because othe damage that can result from the loss of fans, heaters, pumps, etc.

VENTILATION

Ventilation is a continuous process which removes moisture from inside a building and provides fresh air for animals. Ventilation also removes excess heat, especially during hot weather, and removes odors, gases, dust, disease organisms, and other contaminants from the animal environment.

Principles of Ventilation Design

Temperature Control

Ventilating rates are usually expressed in cubic feet per minute, cfm.

Ventilating systems are designed to remove moisture and odor in cold weather and excess heat the rest of the year. To maintain a constant room temperature, the heat **produced** by animals and heaters has to equal heat **lost** through building surfaces and ventilation. In a well insulated building, most of the heat loss is in the ventilating air.
- If heat loss exceeds animal heat production, provide supplemental heat.
- If animal heat production exceeds heat loss, increase ventilation.

Moisture Control

When warm moist air contacts a cold surface, the water vapor condenses into water. If the surface is cold enough, the water freezes. Adequate insulation keeps indoor surfaces warm enough that indoor vapor is less likely to condense.

Prevent moisture problems in sidewalls and ceilings with vapor barriers, which keep water vapor from entering where it may condense into free moisture. Install a vapor barrier on the warm (indoor) side of all insulated building sections. Most building materials are highly permeable and **are not** good vapor barriers.

One of the best vapor barriers for farm structures is polyethylene film—4 mil or thicker. Metals are also good vapor barriers, but be sure to seal the joints between sheets.

Adequate ventilation to exhaust stale moist air, frequent cleaning of livestock buildings, and good maintenance of waterers in buildings with solid floors help reduce the amount of vapor in indoor air. Liquids drain through slotted floors, which also reduces the moisture load on the ventilating system.

During cold weather, ventilation brings cold, relatively dry air into the building. The air is warmed by energy from animals, electrical equipment, and heaters. As the air warms, it can hold more moisture and its relative humidity decreases. This ventilating air picks up moisture from the building and exhausts it from the building at a high relative humidity, Fig 36.

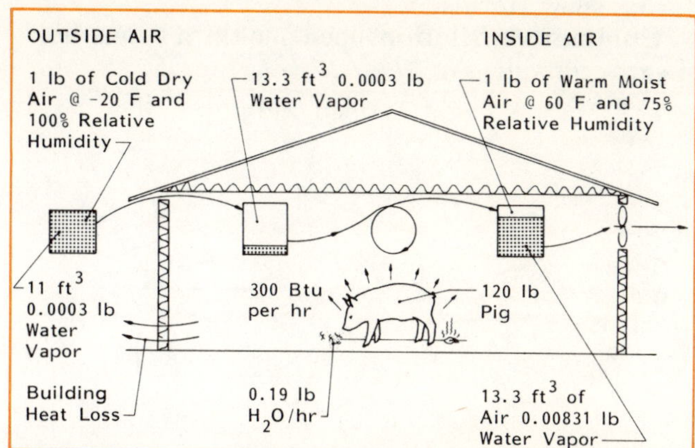

Fig 36. Moisture removed by ventilating systems.

Odor Control

Odor levels depend on type of animal, manure handling system, air temperature, and ventilation rate. The attitude of the manager determines what odor level is acceptable.

Except during the coldest part of the year, ventilating enough to remove excess animal heat and moisture also controls odor. A cold weather ventilating rate that just prevents moisture buildup may not give good odor control.

It is very important that you vent gas, liquid, or solid fuel heaters to the outside with a fire-safe chimney or flue, in order to remove combustion by-products. Or, ventilate at 4 cfm/1000 Btu heater capacity in addition to recommendations in Table 11.

It is also vital to ventilate indoor manure storage pits to reduce gases at the animal level. Exhaust the air required for constant cold weather ventilation through the pit area. Use high volume ventilation while agitating or emptying indoor storage pits, or remove the animals from the building and stay outside yourself.

Natural Ventilation

Natural ventilation can be used for young animals on bedded floors and in well insulated modified-environment buildings with zone heat. Natural ventilation is more suitable than mechanical for mature animals.

Principles of Natural Ventilation

Wind pressure and the difference between inside and outside temperature move air through the building. Natural ventilation works best with small openings at the eaves and ridge of the roof, large sidewall openings for summer, and no ceiling.

Winter ventilation: wind across the open ridge creates suction which draws warm, moist air out through the ridge and fresh air in through the eave openings. Some ventilation still occurs on calm days because warm air rises. Have winter openings at or near the eaves along both sidewalls and at a high point in the roof, Fig 37. Eave opening area is about twice the ridge opening. Close windward openings during snow storms.

Summer ventilation: open one-third to one-half of each sidewall, Fig 37.

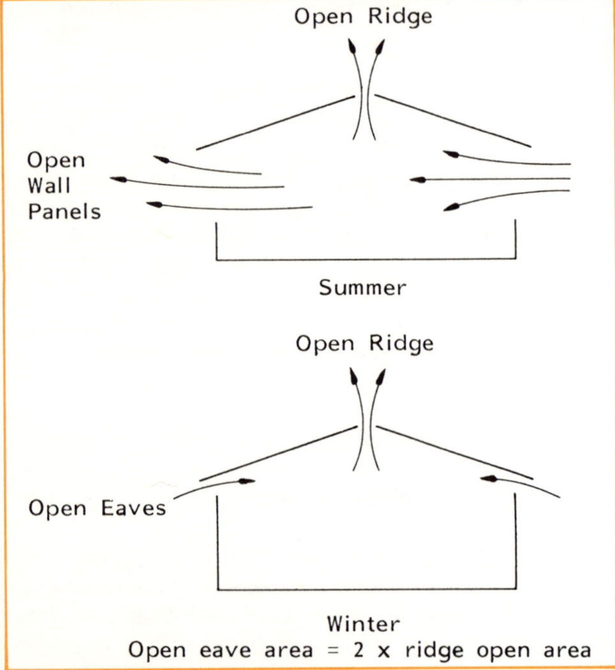

Fig 37. Natural ventilating airflow.

Mechanical Ventilation

A mechanical ventilating system has fans, controls, and air inlets. It provides more control over room temperature and air movement than natural ventilation. It works especially well with young animals because they are susceptible to low temperatures, sudden temperature changes, and drafts.

An effective ventilating system automatically moves the required amount of air through the building and distributes and mixes the air within the building without drafts.

Positive Pressure Ventilation

Fans force fresh air into the building, where it is distributed with a baffle in front of the fan or a duct with carefully-placed outlets, Fig 38. Size the duct with Table 10. Insulate ducts to prevent condensation.

Size positive pressure exhaust outlets at 300 in²/1000 cfm of fan capacity. For example, provide one pair of 2″ diameter holes for each 20 cfm or one pair of 3″ diameter holes for each 45 cfm of airflow through the duct. Base number of holes on the highest airflow through the duct, and block off part of each hole when operating at lower airflow rates. Space holes uniformly along the duct.

Positive pressure systems tend to force moist air into the walls and attic, so a good vapor barrier is important.

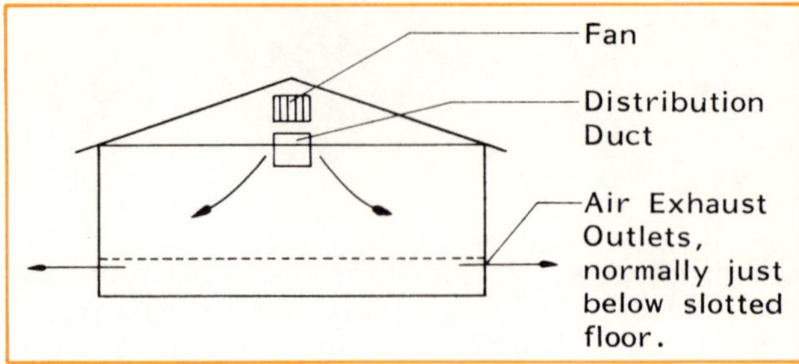

Fig 38. Positive pressure ventilation.

Table 10. Duct sizes.
Minimum clear inside duct dimensions to maintain 600 fpm air velocity.

Airflow rate cfm	Inside duct dimensions Rectangular in. x in.	Round diam, in.
250	6 x 10	9
500	10 x 12	12
750	10 x 18	15
1000	12 x 20	Not practical
1250	15 x 20	
1500	18 x 20	
2000	18 x 27	
2500	18 x 34	
3000	18 x 40	
3500	24 x 35	
4000	24 x 40	
5000	24 x 50	
6000	30 x 48	
7000	36 x 48	
8000	36 x 54	
9000	36 x 60	
10000	42 x 58	
12000	48 x 60	
15000	54 x 68	

Exhaust Ventilating Systems

Exhaust fans force air out of the building, creating a negative pressure, Fig 39, which draws fresh air into the building through baffled inlets. If the inlets are sized and located properly, fresh air is distributed uniformly throughout the room.

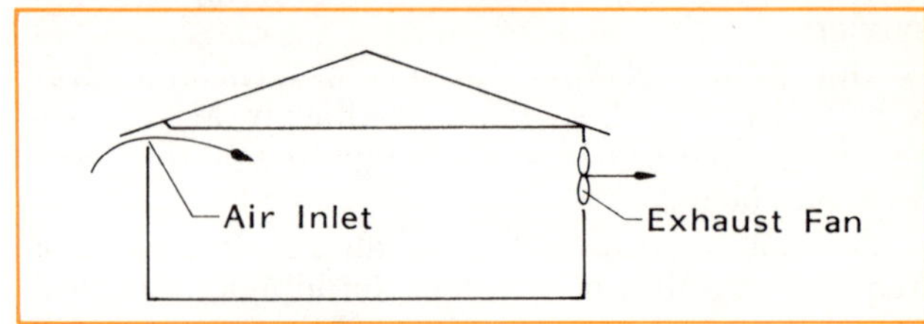

Fig 39. Exhaust (negative pressure) ventilation.

Air Inlets

One of the best inlets is a continuous slot with an adjustable baffle. The baffle restricts the inlet opening to increase air velocity and improve air mixing. Rigid baffles make a uniform slot for even air distribution. Tightly close all doors, windows, and unplanned openings, especially when operating at low ventilating rates. Provide hovers for small animals penned directly under a slot inlet.

Inlet location

For buildings up to 40′ wide, place slot inlets at the ceiling along both sidewalls; for wider buildings, add one or more interior ceiling slots.

Air should travel less than 75′ from inlet to fan. During cold weather, close and seal air inlets within 8′ of the winter exhaust fans.

Inlet size

Size inlets to maintain high air velocity (700 fpm) where the air leaves the baffle. If the velocity is less, cold air settles too rapidly and may chill the animals; if greater, static pressure may decrease fan delivery. Usually, inlets are sized for hot weather, then reduced with a baffle for lower ventilating rates. A 1″ wide, 1′ long slot delivers 50 cfm at 700 fpm. A 2″x1′ slot delivers 100 cfm at that velocity.

Example 13:

Calculate slot openings for a 24′x84′, 30-sow farrowing building. Group exhaust fans in the center of the south wall and locate inlets at the ceiling on the long sides of the room.

1. Total slot inlet length is:
   2 x building length = 2 x 84′ = 168′
   In cold weather, close the inlets over the continuous exhaust fans to prevent short circuiting of air. **Winter** slot opening is about 150′.
2. Maximum ventilating rate per foot of inlet is: (See hot weather rate, Table 11.)
   500 cfm/sow x 30 sows ÷ 168′ = 89.3 cfm/ft
3. At 50 cfm per foot of length for each inch of slot width, hot weather slot width is:
   89.3 cfm ÷ (50 cfm/in) = 1.8″ or about 2″ wide
4. In cold weather, the ventilating rate is much less:
   20 cfm/sow x 30 sows ÷ 150′ = 4.0 cfm/ft
5. For cold weather, partly close the slot with the baffle:
   4.0 cfm ÷ (50 cfm/in) = 0.08″

It is difficult to adjust the baffle to so small a slot opening. Close off every other 4′ section of inlet and double the slot opening in the remaining 4′ sections.
20 cfm/sow x 30 sows ÷ 75′ = 8 cfm/ft
8 cfm ÷ (50 cfm/in.) = 0.16″
Use a ⅛″ slot opening.

Unplanned air leaks make it difficult to maintain adequate air velocity at low ventilating rates. Decrease the inlet opening even more to account for these air leaks.

Cold weather building inlets

Unless you are using a solar collector, bring winter fresh air from the attic to take advantage of natural preheating. Run moisture control fans **continuously,** because when fans are off, warm room air rises through the slot, condenses on the underside of the cold roof, and drips on attic insulation. Use anti-backdraft curtains.

Air enters the attic through gable louvers or ridge ventilators, Fig 41.

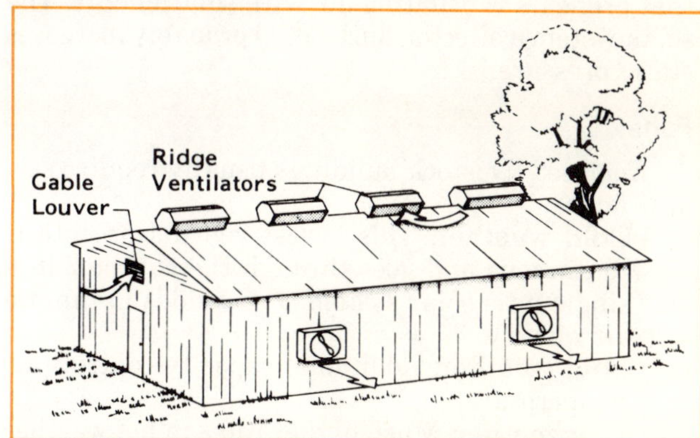

Fig 41. Gable louvers and ridge ventilators.
Required attic opening area, ft² = mild weather ventilating rate, cfm ÷ 200, for summer attic venting and for winter air inlet. Screen attic openings with ½″ hardware cloth.

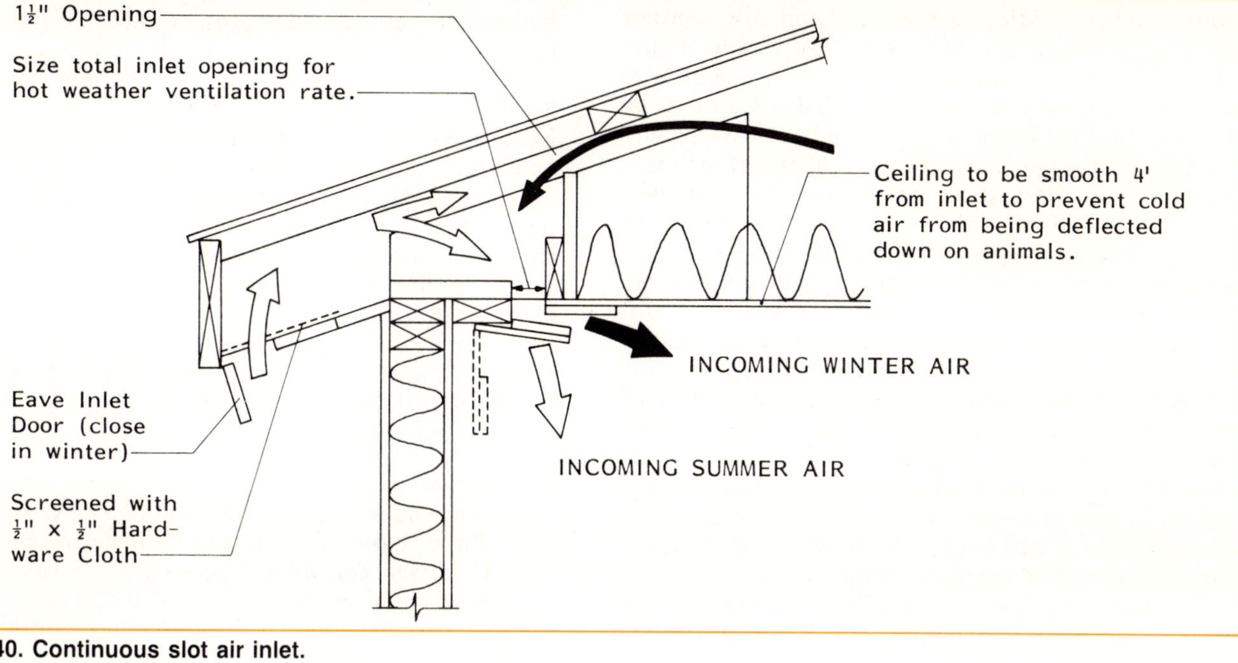

Fig 40. Continuous slot air inlet.
Do not mount pipes or lights within 4′ of inlets.

Hot weather building inlets

In hot weather, pull fresh air directly from the outside instead of the attic—open doors in both eaves, Fig 40. Some of the air entering at the eaves ventilates the attic, lowering the ceiling temperature. Air escapes from the attic through ridge vents and gable louvers. Screen attic openings with ½″ hardware cloth. Smaller mesh screen plugs rapidly with dust and restricts air flow.

Inlet controls

Air inlet control is critical to good ventilation. The size of the slot inlet should change each time the ventilating rate changes, preferably automatically. Operate manually-adjusted baffles from one location with a winch and cable system. Install a manometer, which measures the static pressure, next to the winch for more accurate baffle adjustments. Static pressure is the difference in atmospheric pressure inside and outside the building. A slot width sized to deliver air at 700 fpm creates about 0.04″ static pressure difference. You may need special controls for a system that preheats ventilating air with solar energy. The solar system (collector, ducts, etc.) probably increases static pressure.

Fans

Enclosed livestock buildings usually require several fans:

- **Cold weather.** This lowest ventilating rate is continuous and goes through the collector in a single pass solar system. It provides minimum air quality.
- **Mild weather.** Additional capacity controls air temperature.
- **Hot weather.** Much higher rates in hot weather limit indoor air temperature rise to reduce animal stress.

Occasionally a multi- or variable speed fan can provide both the mild and hot weather ventilating rates. Avoid providing both the cold and mild weather rates on only one fan. As speed is reduced to supply the minimum rate, the fan loses much of its ability to deliver air against static pressure. The result may be little or no ventilation during windy cold weather.

The most important feature for ventilating fans is **capacity** at given **static pressures,** usually specified in cfm vs. inches of water. Cfm/watt is a measure of fan efficiency; consider it after rate and static pressure needs are met. Diameter, speed, horsepower, shape of blade, and other features are less important to you. Ventilation is accomplished with airflow (cfm).

Select a fan capable of moving the required amount of air against at least ⅛″ static pressure. Look at the low-flow characteristics of your minimum-rate fan: it should deliver some air against about ¼″ static pressure to avoid running backwards in high wind. Variable-speed fans have poor pressure ratings at low speeds and may not deliver enough air against wind.

Use fans designed specifically for animal housing and tested against the standards of the Air Movement and Control Association (AMCA). Buy totally enclosed split-phase or capacitor-type fan motors. Protect each fan with a fused switch near the fan, and size the fuse at 25% over fan amperage. Connect each fan to a separate circuit to eliminate shutdown of the entire ventilating system if one motor blows a fuse.

Fan location

With a properly adjusted rigid baffle slot inlet, fan location has little effect on air distribution. If possible, place winter exhaust fans downwind from prevailing winter winds. Space fans so air moves no more than 75′ from inlets.

If fans must exhaust against prevailing winter winds, protect them with a low windbreak 5′-10′ from the building. Fan hoods can reduce wind effects. Put anti-backdraft shutters on all non-continuous fans. Place the shutters on the inside (animal side) of the fan to decrease freezing problems. Space mechanically ventilated buildings at least 35′ apart so fans do not blow foul air from one building into the intakes of another. Do not discharge building air onto a solar collector.

Fan controls

Choose thermostats designed especially for livestock housing with an off-on range of 5 F or less. Contacts must be sealed. Locate thermostat:

- At or near center of building width.
- As low as possible, but out of animal reach.
- Away from cold walls and ceilings.
- Out of the path of furnace exhausts, ventilating inlet air, and direct sunlight.

Determining fan sizes and thermostat settings

Required ventilation varies with outside temperature. Select fan capacities and thermostat settings for a **gradual** airflow increase with an outside temperature increase. Use several fans, multi-speed fans, or variable-speed fans.

Determine fan sizes for positive pressure or exhaust systems from the rates in Table 11 and from product literature.

Most thermostats have an on-off range of about 3-5 F; stage fans to come on in 2-4 F increments. Make sure the heater operates only when ventilation is at the cold weather rate.

Base ventilating rates on animal requirements, not on the solar system flow requirements. Although this may make some solar components less efficient, it is better for the animals and is usually more energy efficient overall. If the solar collector uses separate fans, run solar collector fans and cold weather ventilating fans on separate controls. The cold weather fan must run day and night while the collector fan may run only during the day.

Example 14:

For the 30-sow farrowing building of Example 13, Table 11 gives the following ventilating rates:
Cold weather: 30 x 20 = 600 cfm total
Mild weather: 30 x 80 = 2400 cfm
(Additional capacity = 2400 - 600 = 1800 cfm)

Table 11. Recommended ventilating rates.
The rate for each season is the total capacity needed. Increase rates by 4 cfm/1000 Btu heater capacity if heaters are unvented.

Animal type	Weight lb	Cold weather rate	Mild weather rate	Hot weather rate
		- - - - - -cfm/hd- - - - - -		
Swine				
(For swine, circulation fans can replace exhaust fans for one-half of the hot weather ventilating rates.)				
Sow and litter	400	20	80	500
Prenursery pig	12-30	2	10	25
Nursery pig	30-75	3	15	35
Growing pig	75-150	7	24	75
Finishing pig	150-220	10	35	120
Gestating sow	325	12	40	150*
Boar	400	14	50	300
Dairy	**Unit**			
Cow in warm barn (tie stall or free stall)	1000 lb	25	100	300 to 500
Calf, warm barn	100 lb	10	25	50
Milkroom	cfm			600
Milking parlor	cfm/stall		100	400
Beef				
Cattle in warm confinement	1000 lb	15	100	200 (min.)
Poultry				
Chick	cfm	0.1/bird	0.5/bird	1/lb
Layer, pullet breeder, broiler	cfm/bird	0.50	2	4 (min.)
Turkey poults	cfm/lb	⅙	¼-⅓	

*Use 300 cfm as the hot weather rate for gestating sows in a breeding facility.

Hot weather: 30 x 500 = 15,000 cfm
(Additional capacity = 15,000 - 2400 = 12,600 cfm)

Choose the desired room temperature from Table 12 (e.g. 70 F). Set heater thermostats about 2 F below the desired room temperature (68 F). With a 5 F range on the furnace thermostat, the furnace shuts off at 73 F. Set the thermostat of the smallest mild weather fan about 3 F above the temperature at which the furnace switches off (76 F). Set the other warm weather fans to turn on at increasing increments of 2-4 F (78 F, 80 F, etc.).

If winter fans are variable-speed, adjust the minimum speed to the cold weather rate (600 cfm). Set the fan controller to begin increasing speed at 3-4 F above the furnace-off temperature (76 F). Most variable-speed fans increase speed over a range of about 5 F. Set the other fans to come on in 2-4 F increments, starting at a temperature 5 F higher than the variable-speed fan setting (81 F, 83 F, etc.).

Supplemental Heat

Table 12 lists typical peak supplemental heat requirements for livestock housing. Zone heat for young animals may also be needed. Solar energy can replace some of the fuel needed for supplemental heat. Solar preheating of ventilating air helps dry a building even in mild weather when heaters would not be used. Install full-size heating equipment because solar energy may not be available when you need it.

Radiant heaters

Radiant heaters work well for zone heating. They allow the animal to find its own comfort level by moving closer to or farther from the heater.

Floor heat

Supply floor heat with solar heated warm air, electric resistance cable, hot water pipes, or electric heating coils in fiberglass pads on an existing floor. For heating a floor in winter (e.g. swine nursery), size the collector for 50% (or 25% without heat storage) of the January floor heat needed.

Floor heat evaporates liquids from the floor surface, which increases relative humidity. Arrange resting areas, waterers, and dunging areas so the heated area is seldom wet. Do not heat the floor under the sow in swine farrowing buildings.

Use waterproof insulation under heated floors and no insulation and little bedding on top of them. Place 1″-2″ of concrete between heating elements and plastic insulation. Put perimeter insulation between a heated slab and an outside wall. Concrete above hot water pipes or electric heat cables must be uniformly thick to prevent hot spots.

Table 12. Supplemental heat requirements per animal.
Sized for twice the cold weather ventilating rate in a moderately well insulated building. Additional zone heat may be needed for young animals.
These heat requirements are approximate peak loads for sizing heaters; the Design section shows how to calculate heat needed.

Animal type	Inside temperature, F	Supplemental heat, Btu/hr-hd	
		Slotted floor	Bedded/scraped floor
Swine			
Sow and litter	80	4000	—
	70	3000	—
	60	—	3500
Prenursery pig (12-30 lb)	85	350	—
Nursery pig (30-75 lb)	75	350	—
	65	—	450
Grow/finish pig (75-220 lb)	60	600	—
Gestating sow/boar	60	1000	—

	Supplemental heat, Btu/hr-hd Bedded/scraped floor
Dairy	
Double 4-HB parlor	50,000
-6 or -8 HB parlor	70,000
Milkhouse	10,000 Btu/hr
Calf housing	1,000 Btu/hr per calf
Sheep	
Lambing	400 Btu/hr per ewe Plus heat lamps
Horse	
Warm barn	5,000 Btu/hr per stall

Start heating the floor at least two days before it is needed to allow time for the concrete to warm up.

Electric heat cable usually has a lower installation cost than hot water pipes, avoids freezing problems, and permits individual thermostats at each pen, stall, or group of pens. Electric heat has a higher operating cost than hot water heat.

Unit space heaters

Unit space heaters are vented or unvented heaters inside the building. Their main disadvantage is that they recirculate dusty, wet, corrosive air through the furnace, often resulting in high maintenance.

Air make-up heaters

Air make-up heaters are unvented units mounted outside the building. They heat up only the incoming ventilating air, so they require less service than units which recirculate room air. The heaters modulate fuel flow to the burner for a constant exit air temperature.

Air make-up heaters supply air to the room, so close down the slot air inlet accordingly. Make sure an air make-up heater supplies no more airflow than the capacity of the winter exhaust fan. Air make-up heaters exhaust the products of combustion into the building, which makes them more efficient than vented heaters (about 90% vs. 70%). But water vapor is one of the gases produced (over 1 lb water/lb of propane burned), so ventilation rates must be increased to remove this extra moisture.

A room air distribution system may be needed for adequate warm air circulation.

Heat exchangers

Heat exchangers extract heat from exhausted ventilating air which otherwise would be lost. Most of the heat may come from condensed water during cold weather. Although heat exchangers have a great deal of merit, dust and corrosion from building exhaust air often decrease their efficiency.

Solar heat

Solar heat can provide some of the energy needed. Management with solar preheated ventilating air is the same as with other systems. With space and floor heating, solar energy supplements heat from other sources. With underfloor warm air heating, you must prevent circulating cold air which could cool the floor instead of warming it.

Energy Conservation in Livestock Buildings

When planning to build or remodel animal buildings, place a high priority on energy conservation and improving energy efficiency. Conserve all the energy you can before buying a solar system.

Ventilation Adjustments

Ventilating air is the largest source of winter heat loss in a well insulated livestock building.

The most cost-effective heat savings is through ventilating system management. Select fans carefully, using separate fans for the three basic ventilating needs. Select a simple control system and watch it carefully, especially during the peak heating season. **Do** maintain minimum ventilating rates for moisture control, but **do not** ventilate any more than you have to.

For example, a 200-pig nursery with a 3 cfm/pig ventilation rate requirement has a 600 cfm continuous fan. If temperatures are 80 F indoors and 10 F outdoors, ventilating heat loss is about 15 gal of propane/day, or about three times the heat loss through the ceiling and walls of a well insulated building.

Building Design, Modifications and Scheduling

Building design and animal scheduling significantly affect heat utilization. Keep the building near capacity all winter. Consider practices such as double- and triple-decking nursery and growing pigs even though they may complicate ventilation.

Arrange building space for heat efficiency:
- Maintain recommended animal density.
- Heat small animals with excess heat from large ones in the same room. However, disease transfer can be a significant problem.
- Separate heated space from cold space.
- Divide empty space from full space with an insulated partition.

Building Insulation

The animals are a major source of heat in livestock buildings, so insulate to conserve animal heat. Adequate insulation cuts heat loss through walls, ceiling, and floor, and improves animal comfort by keeping building surfaces warm. See the insulation section.

Insulation and Heat Loss

Building heat is lost or gained by conduction through construction materials, by air currents moving over the inside or outside surfaces, by radiation from warm surfaces, and by exhausting warm air which is replaced by cold outdoor air. Reduce heat losses with insulation and with ventilation management.

Why Insulate?

First, insulate to conserve heat from animals and equipment (Fig 42) to reduce the amount of supplemental heat required. In livestock buildings, the heat saved can evaporate water from the floor or warm additional ventilating air.

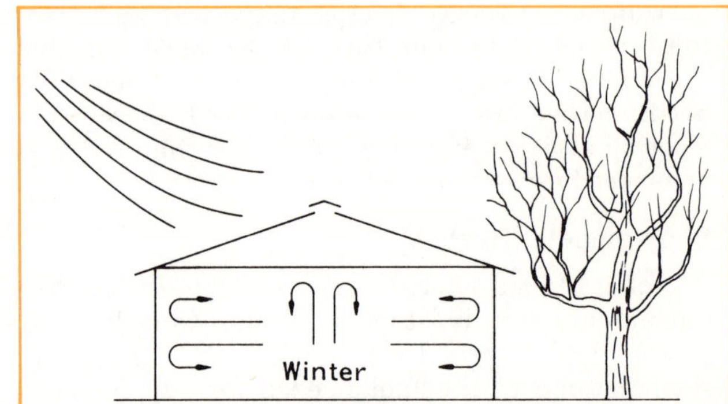

Fig 42. Insulation reduces heat loss during cold weather.

Second, insulate to reduce heat gain in summer, which improves comfort and reduces cooling costs, Fig 43. Wall and roof temperatures of buildings exposed to direct sunlight are as much as 50 F above air temperature.

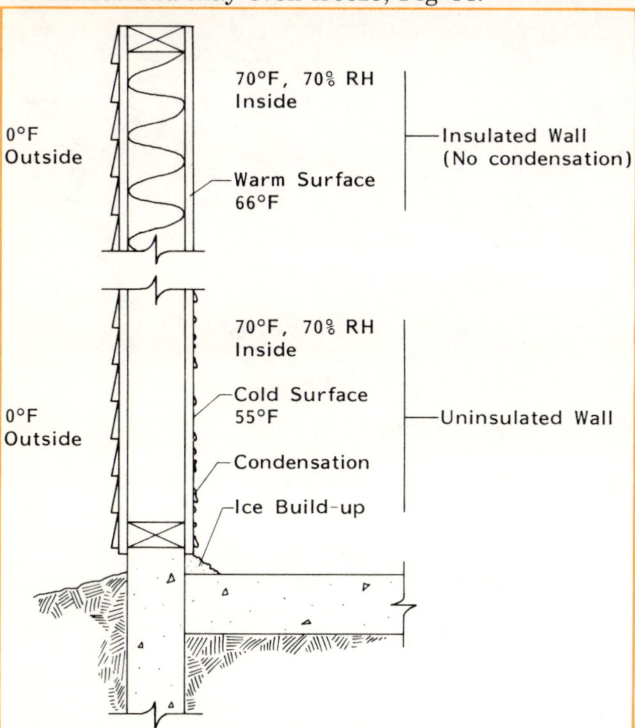

Fig 43. Insulation reduces heat gain during hot weather.

Third, insulate to warm the inside of exterior walls during cold weather. Warmer surfaces reduce condensation or "sweating" and reduce radiant heat loss from objects in the building to cold walls. In poorly insulated livestock buildings, inside ceiling and wall surfaces become cold in winter and moisture condenses and may even freeze, Fig 44.

Fig 44. Warm, moist air condenses on a cold surface.
Insulation helps to control condensation (sweating) by making the wall and ceiling surfaces warmer.

Selecting Insulation

Insulating value varies considerably among different materials. For example, it takes 2.8″ of plywood to equal the insulating value of 1″ of glass wool batt.

Compare insulating materials by their R-value, which measures resistance to heat flow. The higher the R-value, the better the insulation.

Many different insulations are available. See Table 13. Manufacturers report R-values for the insulation alone, or higher values as installed R including air films, etc. Be sure you compare insulation on the same basis, preferably using R-value for the insulation alone.

R-value is additive—2″ of a material has twice the R-value of 1″. Sum R-values for all materials in a given section to get a total R-value.

Table 13. Insulation values.
From 1981 ASHRAE Handbook of Fundamentals. Values do not include surface conditions unless noted otherwise. All values are approximate.

Material	R-value Per inch (approximate)	For thickness listed
Batt and blanket insulation		
Glass or mineral wool, fiberglass	3.00-3.80	
Fill-type insulation		
Cellulose	3.13-3.70	
Glass or mineral wool	2.50-3.00	
Vermiculite	2.20	
Rigid insulation		
Exp. polystyrene,		
extruded, plain	5.00	
molded beads, over 1 pcf	4.20	
Exp. polyurethane, aged	6.25	
Glass fiber	4.00	
Polyisocyanurate	7.04	
Foamed-in-place insulation		
Polyurethane	6.00	
Building materials		
Concrete, solid	0.08	
Concrete block, 3 hole, 8″		1.11
Metal siding, insulation-backed, ⅜″		1.82
Lumber, fir and pine	1.25	
Plywood, ½″	1.25	0.62
Hardboard, tempered, ¼″	1.00	0.25
Insulating sheathing, ²⁵⁄₃₂″		2.06
Gypsum or plasterboard, ½″		0.45
Wood siding, lapped, ½″x8″		0.81
Windows, doors (includes surface conditions)		
Single glazed		0.91
with storm windows		2.00
Insulating glass, ¼″ air space		
double pane		1.69
triple pane		2.56
Wood, solid core, 1¾″		3.03
Metal, polystyrene core, 1¾″		2.13
Floor perimeter (per ft of exterior wall length)		
Concrete, no perimeter insulation		1.23
with 2″x24″ perimeter insulation		2.22
Air space (¾″-4″)		0.90
Surface conditions		
Inside surface		0.68
Outside surface		0.17

Example 15:
A typical wall is shown in Fig 45. See Table 13. Note that the blanket insulation provides more than 80% of the total R-value.

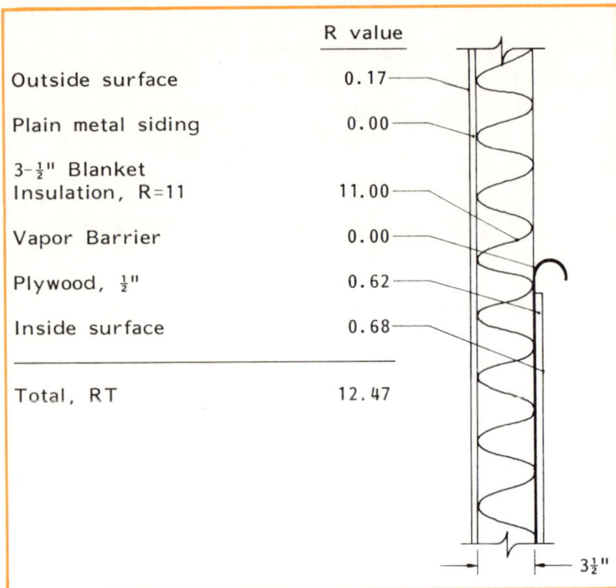

Fig 45. Example 15: R-values for insulated wall.

Important rules for insulating buildings are:
- **Use enough insulation.** Heat loss is directly proportional to the RT value (total R for a section).
- **Rodentproof the building's perimeter.**
- **Insulate all areas.** Insulation on concrete walls and around the building perimeter were often neglected until recent years.
- **Install windows only when necessary.** Single pane glass is poor insulation. If windows are required, use double glazing or storm windows to cut heat loss.
- **Make sure insulation fits snugly.** Fully insulated walls and ceilings have the lowest heat loss. Voids reduce insulating value and add substantially to heat loss.
- **Protect insulation with a vapor barrier.** See Moisture Control.
- **Make structure tight.** Reduce air leaks through cracks with caulking and good construction.

Note: If foam plastic insulations are not protected suitably from potential fire, your insurance company may refuse to provide coverage on the structure.

Recommended Insulation Levels

The amount of insulation needed in farm buildings depends on many factors, such as the expected outside temperature, number and size of animals housed, desired inside temperature, and economics.

Cold environment buildings have indoor conditions about the same as outside conditions. Examples are machinery storages, cold free stall barns, and open-front livestock buildings. Some roof insulation is recommended to reduce solar heat gain in summer and condensation in winter. Open-front buildings may not need insulation to prevent condensation.

Modified environment buildings rely on animal heat to evaporate moisture and maintain the desired inside temperature. They have more insulation than cold environment buildings. Examples are warm free stall barns, poultry production buildings, and swine finishing units.

Supplementally heated buildings are well insulated but require extra heat to maintain the desired inside temperature. Examples are farrowing buildings, farm shops, and offices. Cold and modified environment buildings requiring zone heating, such as floor heat in an open-front building, are **not** classified as supplementally heated.

Determine recommended minimum insulating levels for farm buildings from Fig 46 and Table 14. More insulation may be justified with increasing energy costs in supplementally heated buildings.

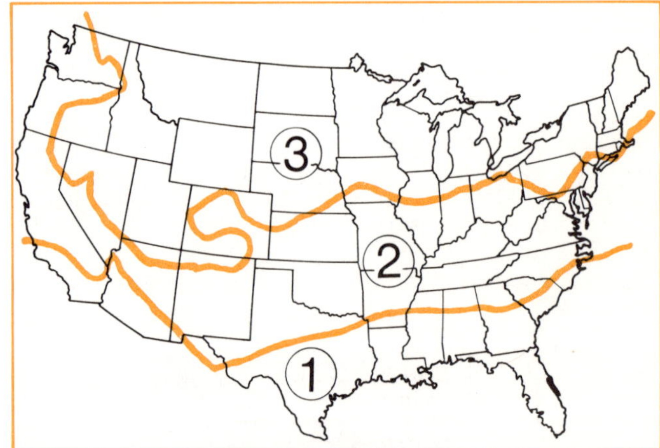

Fig 46. Insulation zones.

Table 14. Minimum insulating levels for farm buildings.
R-values are total for building sections.

Zone	Open-front or cold Walls	Open-front or cold Roof	Modified environment Walls	Modified environment Roof	Supplementally heated Walls	Supplementally heated Ceiling
1	—	6	6	14	14	22
2	—	6	6	14	14	25
3	—	6	12	25	20	33

DESIGN

Solar collectors can provide part of the supplemental heat needed in livestock buildings. This section tells how to determine heat needed and how to size a solar collector.

Calculating Heat Loss

Total heat loss is the sum of the losses from the building envelope (walls, ceilings, foundation perimeter, etc.) and from heating the ventilating air.

Building Envelope

The rate of heat loss through each building component is proportional to its area, the total R-value, RT, for all building components, and the difference between the inside and outside temperatures. The higher the RT value, the lower the rate of heat flow. The heat loss factor, HLF, is the hourly heat loss per degree of temperature difference. Calculate HLF for each building component.

Eq 12.

$$HLF = A \div RT$$

    HLF = heat loss factor for the building component, Btu/hr-F
    A = surface area of component, ft^2
    RT = total resistance to heat flow of the component, $F\text{-}ft^2\text{-}hr/Btu$

The building perimeter is a special case; the RT value, Table 13, is given per foot of length. Replace the area, A, with the length of the exterior wall. Heat loss from the building perimeter is:

Eq 13.

$$HLFp = P \div Rp$$

    HLFp = perimeter heat loss factor, Btu/hr-F
    P = building perimeter, ft, = 2 x (L + W)
    Rp = resistance per ft of building perimeter

Ventilation

If the building is ventilated, calculate ventilating heat loss from:

Eq 14.

$$HLFv = 1.1 \times Qv$$

    HLFv = ventilating air heat loss factor, Btu/hr-F
    Qv = ventilating rate, cfm
    1.1 = a constant to allow for air density and specific heat of air, Btu/F-hr-cfm

For the total heat loss factor for a building, add the factors for each building component, the perimeter, and ventilation. The sample problem illustrates the procedure.

Heat loss per hour depends on the difference between indoor and outdoor temperatures. (Note: this heat loss is **not** the supplemental heat required to maintain building temperature.)

Eq 15.

$$q = HLF \times TD$$

    q = building and ventilation heat loss, Btu/hr
    HLF = total heat loss factor
    TD = temperature difference, F. It is indoor minus outdoor temperature, ti - to, F.

Example—Heat Loss Calculations

Find the heat loss from a 24'x36', 220-pig nursery built as in Fig 47. Assume a 50-lb average pig weight. The inside design temperature is 70 F and the outside design temperature is -10 F. There are two 3'x7' doors, insulated with 1" of molded polystyrene. The building is ventilated at 3 cfm/pig minimum winter rate. Calculations are shown in Table 15.

Step I

List the building dimensions. Calculate the perimeter and the building surface areas. The resistance of the frame wall is 12.47 from Table 13 and Fig 45. The RT value of 2.22 in the perimeter equation assumes 2"x24" perimeter insulation, Table 13.

Step II

Solve the heat loss factor equations with the appropriate values from Step I. Find total HLF.

Step III

Calculate heat loss with Eq 15.

Calculate maximum heat loss at a low winter temperature: -10 F is often assumed in much of the Midwest. Maximum heat loss helps in sizing heaters (Btu/hr capacity).

Calculate average heat loss for each month using the average monthly temperature (Table 27). Size solar collectors and estimate supplemental fuel needs with average heat loss for January.

Steps IV-VII

The example is continued on later pages.

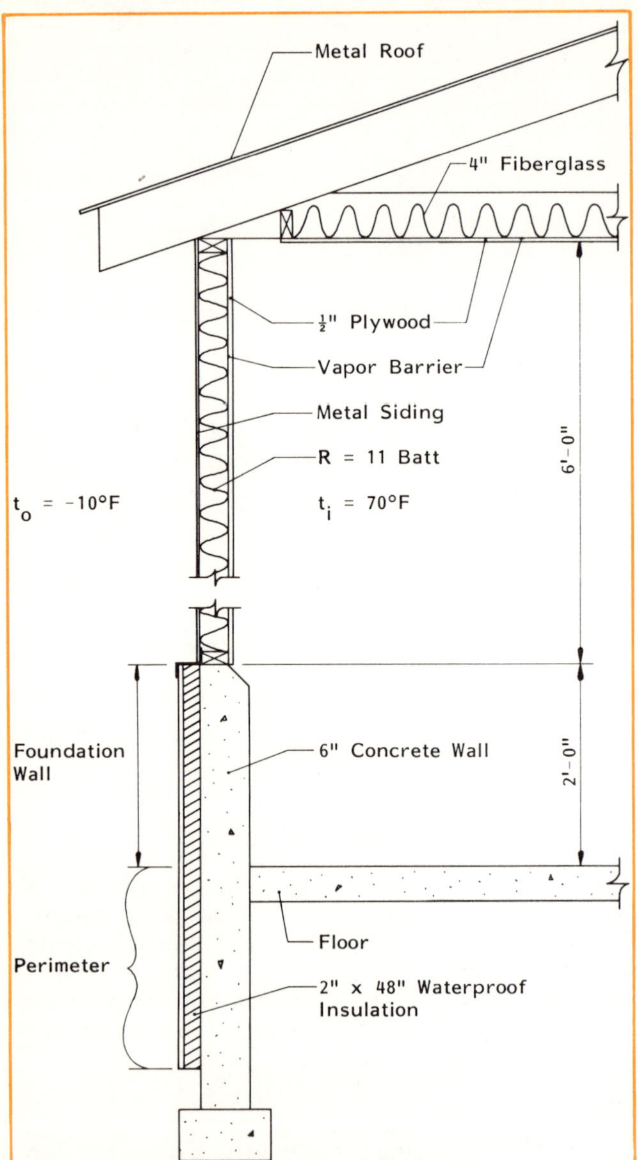

Fig 47. Sample problem building wall.

Supplemental Heat

Because solar heat may not be available when heating need is greatest, install a furnace large enough for the full heating load. Allow for unusually cold weather and the building being empty or only partly full of animals. Calculating expected heat needs ensures adequate capacity for your climate and building.

Because there is usually very little cost increase for substantial increase in furnace capacity(except for electric heat), the recommended heat per animal in Table 12 is generous. Buy a furnace for its rated output, not the input rating quoted in typical sales literature.

Calculate supplemental heat needed to maintain building temperature by subtracting heat produced by the animals from the building and ventilating heat loss.

Eq 16.

$$Q_{sup} = HLF \times TD - AHP$$

Q_{sup} = heat input required, Btu/hr
HLF = total heat loss factor
TD = temperature difference, F. It is indoor minus outdoor temperature, t_i - t_o, F.
AHP = animal sensible heat production, Btu/hr

Animal sensible heat production values are in Table 19. Some animal heat is wasted, because it is produced during the warm part of the day when building and ventilating heat losses are low. To calculate heat needed for a solar-heated livestock building, use 50% of the Table 19 values. More animal heat is utilized if you provide solar heat storage.

Example—Step IV

Calculations for the 220-pig nursery:
HLF = 930 Btu/hr-F (from Step II), t_i = 70 F
From Table 19, q_s = 151 Btu/hr-pig
For 220 pigs and 50% of the animal heat production,
AHP = 151 x 220 ÷ 2 = 16,610 Btu/hr
Use outside temperatures, t_o, for Lemont, Illinois, from Table 27.

For January, for example:
Mean January temperature = 24.8 F
Temperature difference, TD = 70 - 24.8 = 45.2 F
Heat loss, q = 930 x 45.2 = 42,036 Btu/hr
Heat demand, Q_{sup} = 42,036 - 16,610 = 25,426 Btu/hr
Daily heat demand = 25,426 x 24 = 610,224 Btu/day
Size furnace for maximum heat needed, 74,400 Btu/hr, from Step III.

Collector Size

It is seldom practical or economical to provide all supplemental heat with solar energy, especially for collectors without storage, which provide heat only during the day when heating needs are typically lowest. Fig 48 illustrates the effect of collector size on utilization of solar energy. Much of the heat from a large solar collector is wasted in spring and fall, and collector pay-back is slow.

A number of factors and compromises affect final collector size. We suggest you estimate collector size by each of the appropriate methods below, and then compare the results. Seriously consider engaging an engineer experienced in designing solar systems and livestock ventilating systems to help you size your collector.

1. Collector size by physical limitations.
   - Collector size is often limited by the size of the south wall, or south roof slope, of a building.
   - Collector size can be limited by the amount of unshaded area on a building. The collector must be unshaded during the season you need the heat. Avoid trees, silos, and other buildings. See Fundamentals.
   - Portable or other remote collectors need space away from shading, traffic, and ani-

mals. The amount of space available can limit collector size. Ducts get in the way, tend to have high heat loss, leak air, and increase cost, so minimize their length.

2. Collectors sized for minimum winter ventilation. For preheating ventilating air, the maximum collector size is about 1 ft² of collector area for each cfm minimum winter ventilating rate; 0.3 to 1.0 ft² collector area per cfm is a common design range. Collectors are more efficient at higher airflows, so smaller collectors produce more energy per ft².

3. Collector size by heat needs.
   - For heating water, provide 1 ft² collector area per gal of hot water needed per day.
   - For heating a floor in winter (e.g. swine nursery), size the collector for 50% (or 25% without heat storage) of the January floor heat needed. Allow 100 to 130 Btu/hr per ft² of floor area for farrowing and 85 to 100 Btu/hr-ft² for nursery pigs.
   - For preheating winter ventilating air, select collector size to provide:
   Up to 50% of the January supplemental heat needed, with heat storage.
   Up to 25% of the January heat needed, without heat storage.

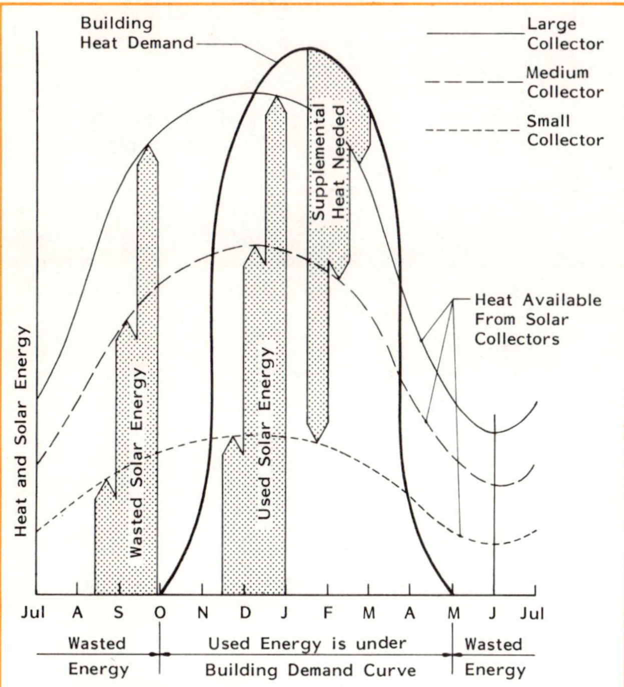

Large Collector
Small proportion of total energy collected is utilized. Collector cost pay-back very slow. Even with large heat storage, furnace is needed for midwinter and for cloudy fall and spring periods.

Medium Collector
Compromise. Much, but not all, furnace operation is replaced with solar energy over several months.

Small Collector
High proportion of available solar energy used. Pay-back rapid. Little effect on furnace operation.

Fig 48. Collector size and solar energy used.
Vertical surface, 40°N latitude.

Table 15. Heat loss calculations.
Data are for sample problem.

Step I

Building dimensions	(ft)
Length (L)	36
Width (W)	24
Frame wall height (H)	6
Concrete wall height (F)	2
Perimeter	120

Surface area	(ft²)
Ceiling area	864
Window area	0
Door area	42
Frame wall area less window & door area	678
Concrete wall area	240

RT values	(Btu/hr-F)
Ceiling	13.98
Window	-
Insulated door	7.42
Frame wall	12.47
Concrete wall	8.50
Perimeter	2.22

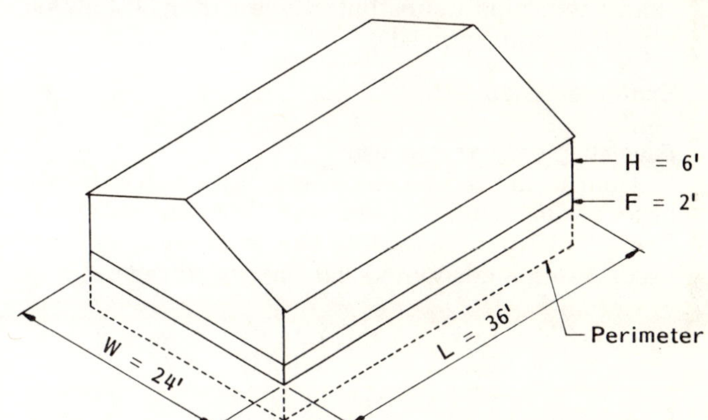

Step II

Item	HLF = Area / RT	= EF
Building		
Ceiling	HLFc = 864 / 13.98	= 61.80
Windows	HLFw = 0 / 1.0	= 0.
Doors	HLFd = 42 / 7.42	= 5.66
Frame walls	HLFw = 678 / 12.47	= 54.37
Concrete walls	HLFf = 240 / 8.50	= 28.24
Perimeter	HLFp = 120 / 2.22	= 54.05
Ventilation	HLFv = 1.1 x vent rate	
	HLFv = .1 x 3x220	= 726.00
	Total HLF	= 930.12
For convenience, round to HLF		= 930 Btu/hr-F

Step III
Heat loss, $q = HLF \times (t_i - t_o)$
Maximum heat needed, at -10 F:
$q = 930 \times [70 - (-10)] = 930 \times 80 = 74{,}400$ Btu/hr
Average January heat need, at 10 F:
$q = 930 \times (70 - 10) = 930 \times 60 = 55{,}800$ Btu/hr

Eq 17.

$$A = 50\% \text{ (or } 25\%) \times Qsup \times 24 \text{ hr/day} \div (RAD \times EFF)$$

A = collector area
$Qsup$ = supplemental heat demand for January, Btu/hr
RAD = average January insolation, Btu/ft^2-day
EFF = collector efficiency

Example—Step V

Sizing collector for preheating ventilating air

Assuming a south sidewall collector with storage and an efficiency of about 30% near Lemont, Illinois, calculate the collector area needed.

January supplemental heat is 25,426 Btu/hr, from Step IV. January insolation from Table 27 is 1159 Btu/ft^2-day on the average for the whole month.

Using Eq 17, collector area is:
$$A = 0.50 \times 25{,}426 \times 24 \div (1159 \times 0.30)$$
$$= 877.52 \text{ ft}^2$$

If the collector is about 8' high, it would need to be 877.52 ÷ 8, or about 110' long. If the collector were at an angle of about 60° instead of vertical, the daily energy would be 1251, instead of 1159, Btu/ft^2-day, which only decreases the needed length to about 102'.

Because the example building is only 36' long and probably limits the total length of the collector, solar energy will supply substantially less than 50% of the expected January needs.

Example—Step VI

Estimating solar energy use

Continuing the previous step, if the sidewall collector is 36' long, 8' high, and 30% efficient near Lemont, Illinois, estimate the total solar energy used. Assume adequate solar energy storage.

From Eq 11:
$$CE = EFF \times A \times RAD$$
$$= 0.30 \times 36' \times 8' \times RAD$$
$$= 86.4 \times RAD$$

Energy used equals energy collected, except that the maximum energy used is the heat demand from Table 16. Only 50%-100% of the solar heat collected will reduce costs.

Table 16. Building hourly heat demand calculations.
Data are for sample problem (Lemont, IL).
Heat demand is heat loss minus 50% of animal heat: Qsup = q - AHP

Month	Mean to	TD = ti-to	q = HLF x TD	Heat demand Qsup	
	F	F	Btu/hr	Btu/hr	Btu/day
Jan	24.8	45.2	42,036	25,426	610,224
Feb	28.4	41.6	38,688	22,078	529,872
Mar	35.6	34.4	31,992	15,382	369,168
Apr	50.0	20.0	18,600	1,990	47,760
May	59.0	11.0	10,230	-	-
Jun	69.8	0.2	186	-	-
Jul	73.4	-	-	-	-
Aug	71.6	-	-	-	-
Sep	64.4	5.6	5,208	-	-
Oct	53.6	16.4	15,252	-	-
Nov	39.2	30.8	28,644	12,034	288,816
Dec	28.4	41.6	38,688	22,078	529,872

Table 17. Daily heat use for each month.
Data for sample problem.

Month	Insolation (Table 27) Btu/day-ft^2	Solar heat collected	Qsup req'd (Table 16)	Daily heat use Solar	Fossil
		- - - - - - - -Btu/day- - - - - - - -				
Jan	1159	100,137	610,224	100,137	510,087
Feb	1144	98,841	529,872	98,841	431,031
Mar	1110	95,904	369,168	95,904	273,264
Apr	876	75,686	47,760	47,760	-
May	830	71,712	-	-	-
Jun	800	69,120	-	-	-
Jul	812	70,156	-	-	-
Aug	953	82,339	-	-	-
Sep	1116	96,422	-	-	-
Oct	1169	101,001	-	-	-
Nov	908	78,451	288,816	78,451	209,970
Dec	901	77,846	529,872	77,846	452,026

For seasonal use of solar energy, multiply the average daily use for each month by the number of days in that month. Table 17 shows calculations for each month. Table 18 shows calculations for each month and the seasonal total.

In January, for example: 31 days x 100,137 and 510,087 (solar and fossil heat use from Table 17) = 3,104,247 and 15,812,697 Btu/month.

Table 18. Seasonal solar energy use.
Data for sample problem.
The ratio of solar to total energy used, 15,044,375 / 71,709,030 = 21%, which is called the solar fraction.

Month	Days	Energy needed Btu/month Solar	Fossil
Jan	31	3,104,247	15,812,697
Feb	28	2,767,548	12,068,868
Mar	31	2,973,024	8,471,184
Apr	30	1,432,800	-
May	31	-	-
Jun	30	-	-
Jul	31	-	-
Aug	31	-	-
Sep	30	-	-
Oct	31	-	-
Nov	30	2,353,530	6,299,100
Dec	31	2,413,226	14,012,806
SEASON	Btu	= 15,044,375	56,664,655
TOTAL	Btu	= 71,709,030, solar + fossil	

Miscellaneous Design Data

• Storages:
  -For liquid systems, allow about 25 lb water (about 3 gal) per ft^2 of collector.
  -For air systems, allow about 1 ft^3 or 100 lb rock per ft^2 of collector. At 1 to 2 cfm/ft^2 of collector, airflow through storage will be about 1 cfm/ft^3 of rock, which gives a 12-hr lag between time of maximum energy at the collector and when that energy is apt to be released to the ventilating air. One ft^3 rock provides about 20 to 30 Btu storage per degree F temperature difference in storage. Expect about a 30 F temperature rise in storage

Table 19. Animal heat production.
ASHRAE values.

Livestock	Temp. ti (°F)	Latent heat, qL		Sensible heat, q_s
		lb H_2O/ pig-hr	- - -Btu/pig-hr- - -	
10-lb pig	85	0.03	18	130
20-lb pig	75	0.04	45	95
30-lb pig	65	0.07	70	165
50-lb pig	40	0.12	126	304
(solid floors)	50	0.13	137	243
	60	0.145	152	208
	70	0.18	189	151
	80	0.235	247	83
100-lb pig	40	0.14	147	443
(solid floors)	50	0.15	158	352
	60	0.18	189	281
	70	0.22	231	199
200-lb pig	40	0.20	210	650
(solid floors)	50	0.21	220	520
	60	0.225	236	414
	70	0.265	278	322
		lb H_2O/ lb-hr	- - -Btu/lb-hr- - -	
Leghorn	33	0.0020	2.1	6.9
(night)	54	0.0024	2.5	5.5
	64	0.0022	2.3	5.3
	82	0.0034	3.5	3.8
(day)	35	0.0022	2.3	8.4
	54	0.0032	3.3	6.6
	63	0.0034	3.5	6.6
	72	0.0034	3.6	6.5
	82	0.0041	4.3	5.9
Turkeys 1 lb	75	0.0059	6.2	10.8
2 lb	75	0.0025	2.6	9.5
		lb H_2O/ cow-hr	- - -Btu/cow-hr- - -	
1000-lb dairy cow	30	0.77	800	2950
	40	0.91	950	2650
	50	1.05	1100	2300
	60	1.28	1340	1900
	70	1.34	1400	1700.

temperature during a sunny day; 30 F rise is 600 to 900 Btu stored per ft² of collector.
• Design airflow through the collector for the minimum ventilation needs of the building. Recommended airflow is 1 to 3 cfm/ft² of collector.

When Not Enough Solar Energy

Solar energy probably will not supply all the supplemental heat needed. Reasons include not enough collector and/or storage, extended sunless periods, or extended or extreme cold periods. It is not economic to totally solar heat most applications.

Solutions include:
• Provide supplemental heat, which is the only acceptable solution.
• Allow room temperatures to fluctuate with outside temperatures. Beware of freezing and causing stress on livestock.

Beware of reducing ventilation rates to save heat. Recommended rates are based on years of experience and research and are intended to maintain production and animal health. Problems resulting from underventilating include:
• Condensation, which corrodes equipment and electrical and electronic components. It can increase respiratory stress and bacterial growth.
• Increased concentration of airborne diseases.
• Too little oxygen (asphyxiation) and/or too much harmful gas (ammonia, hydrogen sulfide, CO, CO_2, etc.).

Economic Justification

Many factors influence how economical your solar energy system will be. Some of these factors are:
• **Collected solar energy** is energy obtained at the collector.
• **Solar energy storage** saves some collected energy for later use. Without storage, energy can be wasted during periods of peak solar collection and no solar energy is available for nights and cloudy days.
• **Utilized collected energy** is the proportion of collected energy actually used. Higher room temperatures and higher ventilation rates increase utilization of solar energy. An unused building has zero utilization. The more solar energy collected, relative to heating needs, the lower will be the portion of that energy that is utilized, Fig 48.

Utilized solar energy substitutes for fossil fuel normally used to heat the livestock building. The money saved on conserved fuel pays for the solar energy collection system. An efficient, well-sized, and well-maintained, low-cost solar system can be paid for in a reasonable time—perhaps 5 to 7 years.

Income from a Solar Collection System

The total energy saved per season is the solar energy used, from Table 18. To find the value of that energy, convert your cost for conventional fuel replaced by solar energy to cost per British thermal unit (Btu). In Eq 18, enter your fuel cost in \$/gal, \$/kwh, etc., and the usable energy from column 4 of Table 20.

Eq 18.

$$EV = FC \div UE$$

EV = energy value, \$/Btu
FC = your fuel cost, \$/unit
UE = usable energy, Table 20

If your system is multiple use (e.g. grain drying or summer water heating), include savings from that application. See Example—Step VII.

For example, if electricity costs \$0.06/kwh:
$$EV = 0.06 \div 3413$$
$$= 0.000018 \text{ \$/Btu (or \$18/million Btu)}$$

Table 20. Usable energy for four fuels.

Values for natural and LP gas and fuel oil are for vented heaters. Although usable energy for unvented heaters is relatively high, the moisture they release into the building must be removed by ventilation.

Fuel type	Total energy	Combustion efficiency	Usable energy UE
LP gas	93,000 Btu/gal	0.80	74,400 Btu/gal
Natural gas	100,000 Btu/ccf	0.75	75,000 Btu/ccf
Fuel oil	138,000 Btu/gal	0.65	89,700 Btu/gal
Electricity	3,413 Btu/kwh	1.00	3,413 Btu/kwh

Example—Step VII

Estimate the money solar energy will save on fossil fuel. In Eq 19, multiply the solar energy per season, Table 18, by the energy value, Eq 18:

Eq 19.

$$MS = SE \times EV$$
$$= 15,044,375 \times \$0.000018/Btu = \$271/season$$

MS = money saved, \$/season, energy saved, x energy value
SE = solar energy used, Btu/season, Table 18
EV = energy value, Eq 18

If the money saved is less than the annual cost of the solar collection system, solar collection will not pay for itself. Estimate the annual cost of your collection system as outlined below.

Increase return on solar investment by increasing your use of the solar equipment. Consider other applications for the heat from your solar system. See the *Low Temperature & Solar Grain Drying Handbook,* MWPS-22, for a discussion of solar grain drying. Note that it is difficult to design a solar energy collection system that works well for both high airflow/low temperature (e.g. grain drying) and low airflow/high temperature (e.g. home heating) applications.

Estimating Solar System Costs

Table 21 is a convenient form for summarizing your cost data.

First cost

First cost is the extra investment needed to collect and use solar energy. Include everything extra needed for solar: collector, vents, ducts, storage, fans, controls, shutters, wiring, caulking, fasteners, labor, etc. Deduct any applicable investment and energy tax credits—check with a tax specialist.

For estimating costs before construction, use plans and specifications and talk to those who have built solar systems. Note that first cost is the cost of adding a solar collector to a conventional building. Solar glazing, for example, may also serve as siding; consider only the cost difference over normal siding.

Yearly fixed cost

Annual capital recovery allows for the principal and interest payments on money used to pay for the system.

Eq 20.

$$CR = [(Vf - Vs) \div SPWF] + (Vs \times i)$$

CR = capital recovery, \$/yr
Vf = first cost of the solar system, \$
Vs = salvage value of the system after "N" years, \$. Can be zero.
$SPWF$ = series present worth factor, Table 22
i = interest rate paid on borrowed, or earned on saved, money

Most of the components of a solar collection system should last 10 years, which is a reasonable value for N. Because polyethylene covers last only a year, add their replacement cost to the annual repair cost.

Table 21. Cost summary.

First cost

1. Materials		_______
2. Additional equipment (Ducts, controls, etc.)	+ _______	
3. Labor	+ _______	
4. Total construction cost	=	_______
5. Less tax credits	- _______	
6. First cost:	=	_______

Yearly fixed costs
Expenses

7. Capital recovery	_______	
8. Property taxes, if any	+ _______	
9. Insurance	+ _______	
10. Total expenses	=	_______

Tax savings

11. Line 10 x tax bracket	_______	
12. Yearly fixed cost (10 - 11)	=	_______

Yearly variable costs
Expenses

13. Repairs, maintenance	_______	
14. Energy to operate, if any	+ _______	
15. Total expenses	=	_______

Tax savings

16. Line 15 x tax bracket	_______	
17. Yearly variable cost (15 - 16)	=	_______

Total Net Yearly Cost (12 + 17) = _______

Example:

A $10,000 solar system will be worth about $2,000 after 10 years. At 10% interest, the annual capital recovery cost is (Eq 20):

$$CR = [(\$10,000 - \$2,000) \div 6.145] + (\$2,000 \times 0.10)$$
$$CR = [1,302] + (200) = \$1,502/yr$$

Real estate and property tax information is available locally. Some states exempt solar installations from property taxes. You can roughly estimate these taxes at 1% of first cost.

Insurance information is available from your agent. Some companies do not insure roof-mounted liquid collectors. You can roughly estimate insurance costs at 1% of first cost.

Tax savings can be estimated by multiplying your annual expenses, which are usually deductible, by your income tax bracket. Tax savings apply to use in agricultural buildings but not in a home.

Yearly variable costs

Repairs and maintenance can be estimated at about 2% of first cost. Add the replacement cost for polyethylene and other short-lived materials.

Extra energy is required if extra fans are needed to run the solar system. Estimate the cost of the extra energy.

Eq 21.

$$FEC = W \times HS \times FC \times 0.024$$

$$
\begin{aligned}
FEC &= \text{fan energy cost, \$/season} \\
W &= \text{watts used by fan} \\
HS &= \text{heating season, days/year} \\
FC &= \text{cost of electrical energy, \$/kwh} \\
0.024 &= 24 \text{ hr/day} \div 1000 \text{ watts/kw}
\end{aligned}
$$

Example:

If a fan draws 165 watts, runs 150 days/yr, and electricity costs $0.06:

$$FEC = 165 \times 150 \times 0.06 \times 0.024$$
$$= \$35.64 \text{ fan operating cost per heating season}$$

Net yearly cost

Net yearly cost is the total annual cost of owning the solar system and will be as accurate as you were when estimating costs and expenses. If this cost is less than fuel savings, Eq 19, solar collection will pay for itself within the assumed 10-yr life of the system.

Annual fuel cost

You can estimate fossil fuel cost. First calculate energy value, EV, with Eq 18.

Eq 22.

$$FF\$ = FFU \times EV$$

$$
\begin{aligned}
FF\$ &= \text{fossil fuel cost} \\
FFU &= \text{fossil fuel used, Btu/season, Table 18} \\
EV &= \text{energy value Btu/unit, Eq 18}
\end{aligned}
$$

Table 22. Series present worth factors (SPWF).
Factors for computing annual cost of investment over "N" years of life at the interest rates shown.

N	6%	8%	10%	12%	14%	16%	18%	20%	N
1	0.943	0.926	0.909	0.893	0.877	0.862	0.847	0.833	1
2	1.833	1.783	1.736	1.690	1.647	1.605	1.566	1.528	2
3	2.673	2.577	2.487	2.402	2.322	2.246	2.174	2.106	3
4	3.465	3.312	3.170	3.037	2.914	2.798	2.690	2.589	4
5	4.212	3.993	3.791	3.605	3.433	3.274	3.127	2.991	5
6	4.917	4.623	4.355	4.111	3.889	3.685	3.498	3.326	6
7	5.582	5.206	4.868	4.564	4.288	4.039	3.812	3.605	7
8	6.210	5.747	5.335	4.968	4.639	4.344	4.078	3.837	8
9	6.802	6.247	5.759	5.328	4.946	4.607	4.303	4.031	9
10	7.360	6.710	6.145	5.650	5.216	4.833	4.494	4.192	10
11	7.887	7.139	6.495	5.938	5.453	5.029	4.656	4.327	11
12	8.384	7.536	6.814	6.194	5.660	5.197	4.793	4.439	12
13	8.853	7.904	7.103	6.424	5.842	5.342	4.910	4.533	13
14	9.295	8.244	7.367	6.628	6.002	5.468	5.008	4.611	14
15	9.712	8.559	7.606	6.811	6.142	5.575	5.092	4.675	15
16	10.106	8.851	7.824	6.974	6.265	5.668	5.162	4.730	16
17	10.477	9.122	8.022	7.120	6.373	5.749	5.222	4.775	17
18	10.828	9.372	8.201	7.250	6.467	5.818	5.273	4.812	18
19	11.158	9.604	8.365	7.366	6.550	5.877	5.316	4.843	19
20	11.470	9.818	8.514	7.469	6.623	5.929	5.353	4.870	20

SOLAR APPLICATIONS

This section describes some applications of solar energy to livestock buildings. The systems and installations are, or are very similar to, ones that have been tested in the North Central Region. The efficiencies given are the best available or the best estimate of the authors.

Illinois Portable Collector

Developed at the University of Illinois
by D.W. Morrison and W.H. Peterson

This collector is an example of preheating ventilating air in a one pass system with no storage, Fig 24.

Description

This 12'x24' portable collector was designed for heating air for ventilating livestock buildings or farm shops and for fall grain drying. By moving the collector, greater use of solar energy can sometimes be made without building expensive and heat-wasting ducts.

For ventilating livestock buildings, outside air enters the upper chamber, is drawn down behind the absorbing plate, and is ducted from the lower chamber to the building. For home or shop heating, operate the collector in a recirculating mode (Fig 28) by bringing return air into the lower chamber. For grain drying, draw outside air into the upper chamber and also through the inlet door at the top of the collector plate.

The 63° tilt angle is suitable for fall and winter solar energy collection at latitudes near 40°. By adjusting the length of the vertical studs, the tilt angle can easily be changed for your location.

The top-of-floor to top-of-plate distance on back of a collector for the desired tilt angle should be:

45°	3'- 4½"
50	4'- 0¼"
55	4'- 8¾"
60	5'-10¼"
65	7'- 2¾"

Cover endwalls, back, and floor of the collector with plywood and insulate as shown, Fig 50. Duct heated air through an opening in the end or back of the collector. Anchor securely against wind damage.

Performance

Efficiency of this collector is about 30%. Calculated performances are in Tables 23, 24, and 25.

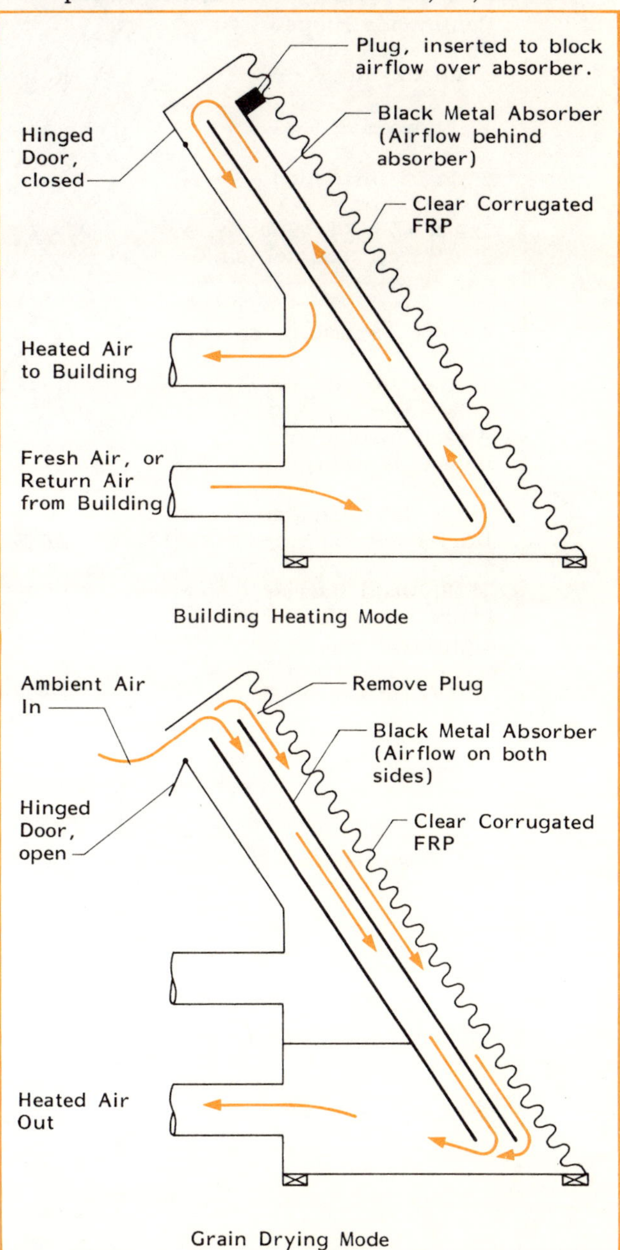

Fig 49. Airflow through a portable collector.

Table 23. Portable collector performance—50 F return air.
Estimated performance for building heating with 50 F return air and 20 F outside temperature.

Airflow cfm	Maximum temperature out, F	Average daily energy collected Btu/day		
		October	December	February
250	123	89,900	63,200	66,700
500	103	130,400	94,800	98,300
1000	85	175,300	126,400	133,400
1500	77	202,300	143,900	151,000
2000	72	224,800	161,500	165,000

Table 24. Portable collector performance—65 F return air.
Estimated performance for building heating with 65 F return air and 20 F outside temperature.

Airflow cfm	Maximum temperature out, F	Average daily energy collected Btu/day		
		October	December	February
250	130	76,400	49,200	58,400
500	112	107,900	70,200	85,600
1000	98	148,400	98,300	116,700
1500	89	170,800	115,900	136,200
2000	85	188,800	129,900	151,700

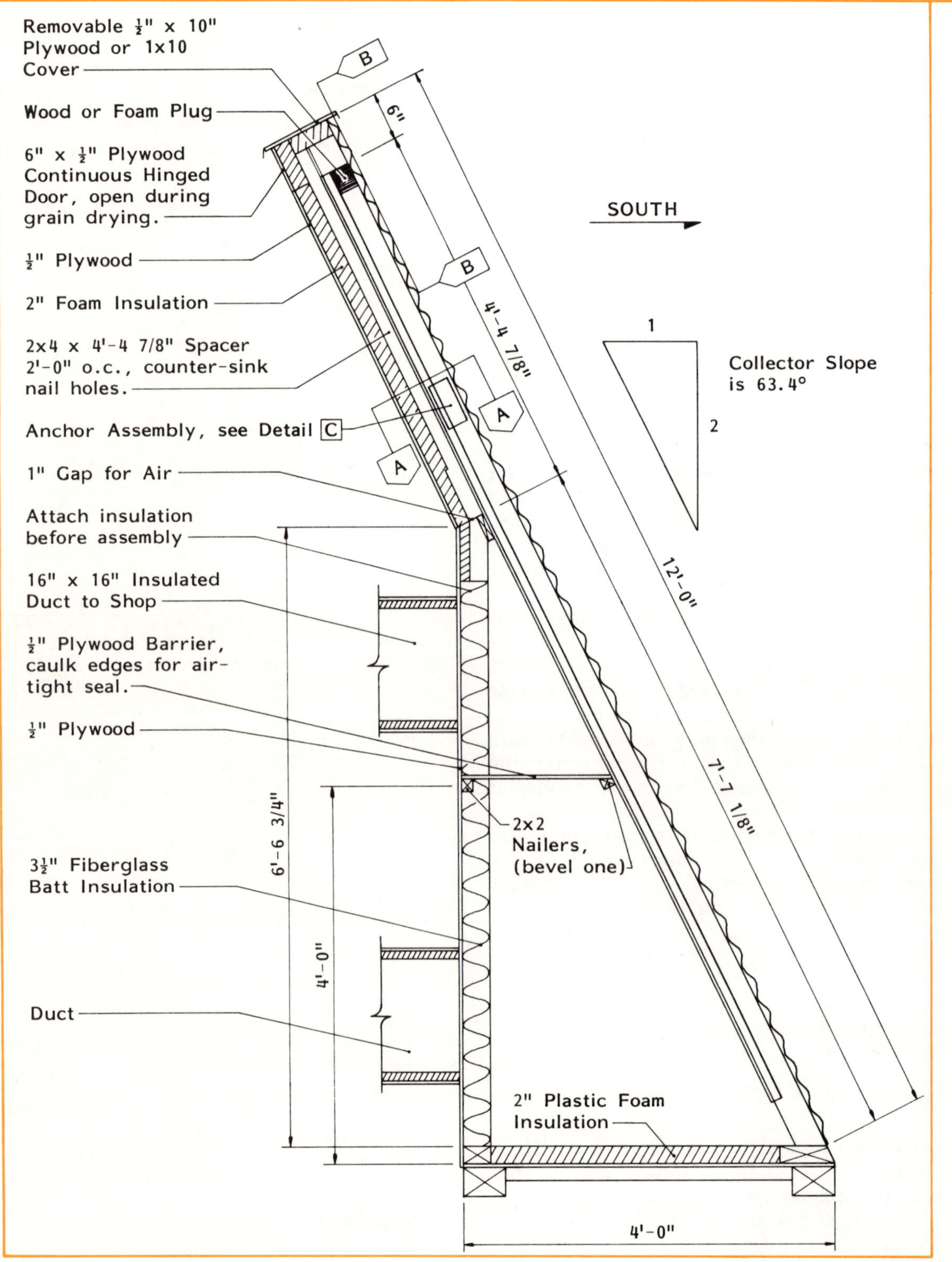

Fig 50. Illinois portable collector plan.

Continued on next page.

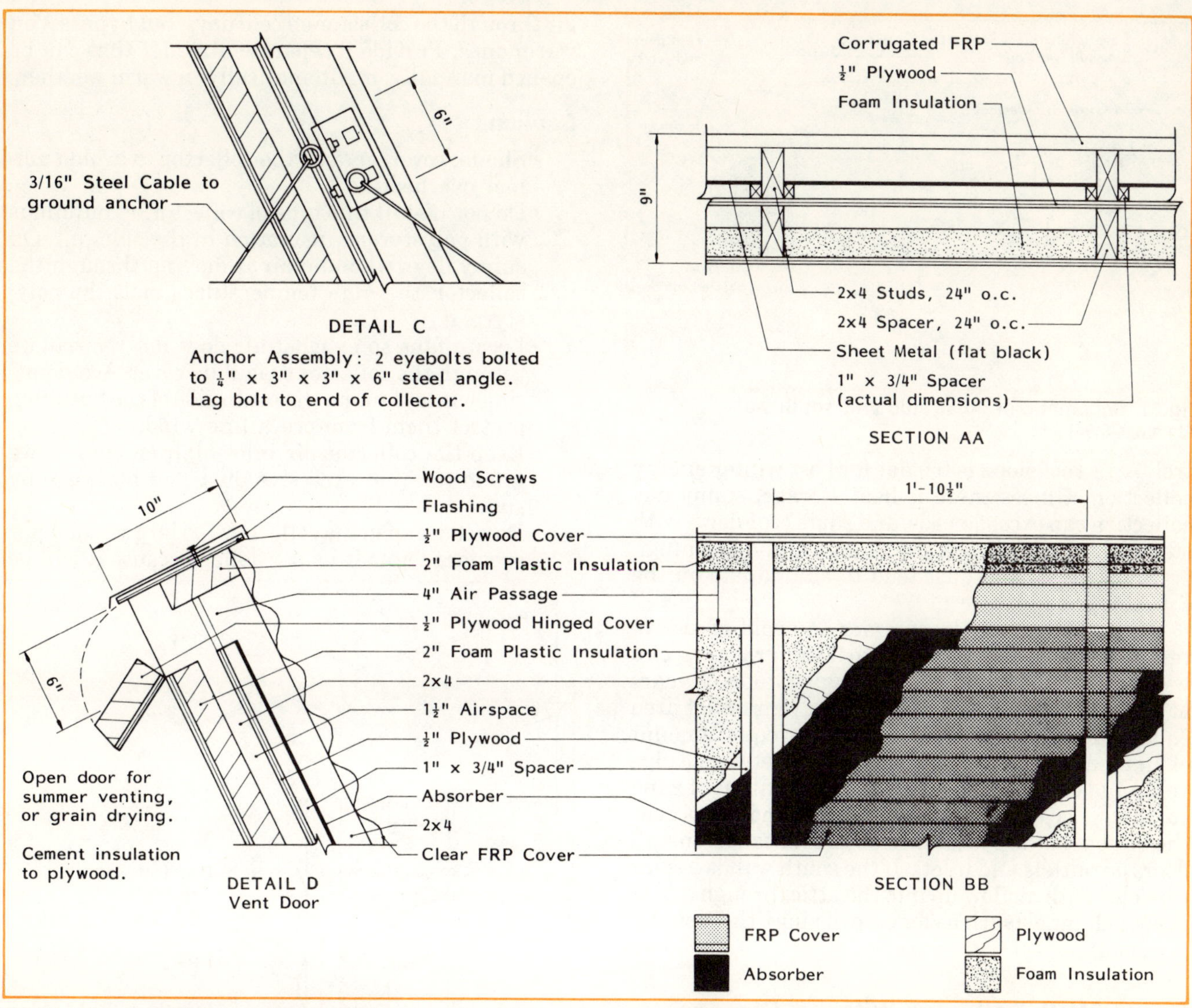

Fig 50. Illinois portable collector plan.

Table 25. Portable collector performance—grain drying.
Estimated performance for grain drying mode.

Airflow cfm	Maximum temperature out, F	Average daily energy collected Btu/day		
		October	December	February
500	77	215,800	168,500	186,800
1000	48	274,200	214,200	237,300
2000	28	310,200	242,200	268,500
4000	15	341,700	266,800	295,700
6000	10	350,700	273,800	303,500
8000	8	355,200	277,300	307,400

Illinois Solar Attic

Description by W.H. Peterson, University of Illinois

This collector is another example of a one pass system with no storage. If rock heat storage is included, it is also an example of the system in Fig 25.

Description

A south-facing roof slope can incorporate a solar collector. In its simplest form, roofing metal is re-placed with a collector cover material, typically FRP. A vertical partition is installed under the ridge of the roof and painted black on the south side. The partition blocks off the north half of the attic to reduce heat loss. Install a ceiling under the truss bottom chord and paint the top side black. One-half of the attic becomes a solar collector.

Air can be circulated from the attic space to the space below or to rock heat storage, or can be ducted to a grain drying system. For space heating, draw air from near the peak, where it is warmest.

Application of this system to a machinery shed and shop, along with optional use of the heat for grain drying, is available in plan mwps-81901, 48′ Solar Machine Shed and Shop.

Advantages and Disadvantages

Installation costs are typically low, and a large area is usually available for the solar collector.

Efficiency of attic units is also typically low in livestock applications—about 15%. It can be 50% for drying applications. Air leakage is difficult to con-

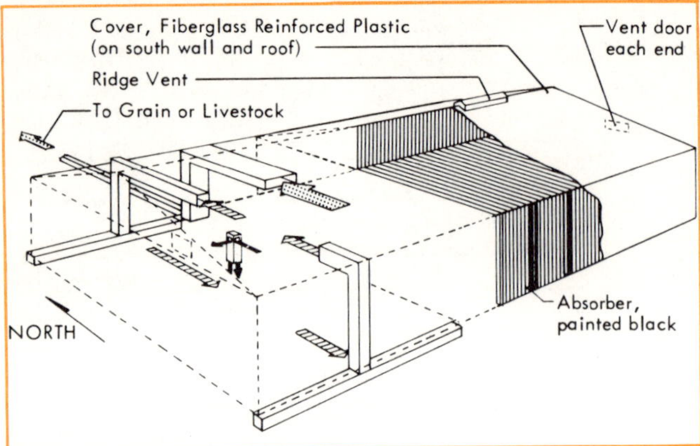

Fig 51. Schematic of solar attic and south wall. Plan mwps-81901.

trol. A 4/12 roof slope is too flat for best winter energy collection. Unless there is heat storage, sunny day collection can exceed needs, and rapid cool-down with sunset does not provide evening or night warming.

Snow, frost, and dust tend to accumulate on the collector surface.

Vent solar attics in summer to avoid excessive temperatures. Frequent high temperatures can weaken framing members, although it takes 50 years at 190 F to cause ignition. Provide a vent outlet area equal to 1% of the collector area and an equal amount of inlet area. Vents can be divided between ridge vents and end wall louvers to take advantage of wind currents. Locate outlets near the peak and inlets near the bottom of the space, such as in overhangs. Install doors on outlets and inlets. If the south wall is a solar collector with airflow up into the attic through a 6″ or deeper channel (such as for crop drying), this usually is enough.

South Wall Collector Without Storage

Description by W.F. Wilcke, Iowa State University

This collector is another example of a one pass system without storage, Fig 24.

Description

These simple collectors can easily be incorporated into new buildings or added to existing east-west structures. Collector construction involves making openings at the top of the south wall, attaching vertical furring strips, painting the south wall and furring strips flat black (two coats), and fastening transparent cover material. Outside air enters a screened inlet (½″ hardware cloth) along the base of the collector, moves upward between the cover material and the black wall, and enters the building through the openings at the top of the wall. Air is moved by the building exhaust fans and distributed by an appropriate inlet system (see Ventilation section).

Because this collector has no heat storage, temperature rises are high (compared with the block wall types), but occur only when the sun is shining. Therefore, collectors without storage tend to overheat buildings on warm sunny days. Pull all ventilating

air through the collector on cold days, but bypass it on warm ones. Provide a separate air inlet that can be opened manually or automatically in warm weather.

Cautions

- Shade, cover, or vent the collector to avoid summer overheating.
- Do not install this type of collector on buildings with polystyrene insulation in the sidewall. On sunny days when no air is moving through the collector, the high temperatures melt the polystyrene.
- Locate fans so exhaust air does not recirculate through the collector or foul its cover. Avoid putting winter fans on the south side of the building; protect them from prevailing winds.
- Keep the collector air inlet high enough above grade to avoid excessive dust and blockage by snow.
- Do not oversize the collector. Too large a collector provides fewer Btus per dollar because it will be

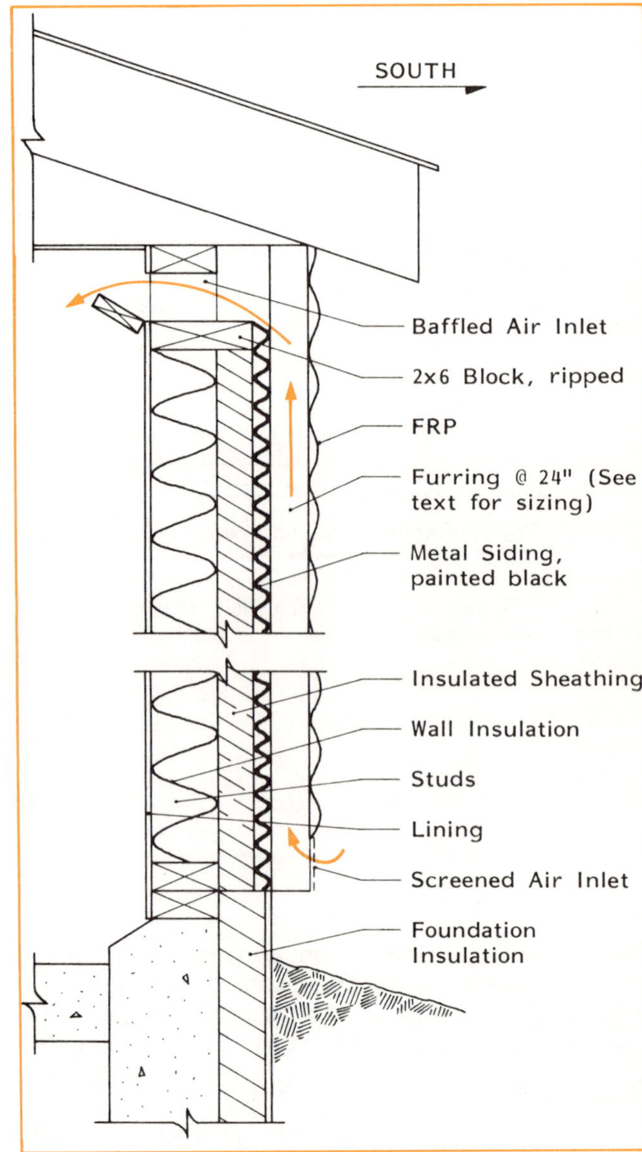

Fig 52. South wall collector.

bypassed on many warm sunny days. Limit collector size to about ½ ft² of collector per cfm, based on minimum winter ventilation. In a typical two-row farrowing house, the 2 cfm/ft² restriction limits collector area to about half the south wall.

Construction Tips

- Size furring strips for an air velocity between 500 and 1000 fpm through the collector at minimum winter ventilating rates. Use the following equation to calculate air channel depth, and then select the commercially available strip (1x2, 2x2, 2x4, etc.) with the closest thickness. Calculated thickness is often very small for farrowing houses, which have low ventilating rates. The smallest furring strip recommended is 1x2, however.

$$FST = cfm \times 12 \div (750 \times CIL)$$

$\quad$ FST = furring strip thickness, in.

$\quad$ cfm = airflow through the collector, cfm

$\quad\quad$ 12 = in/ft

$\quad\quad$ 750 = maximum air speed, fpm

$\quad$ CIL = collector inlet length, ft

- Use a corrugated metal absorber. Other siding materials work, but metal is recommended.
- Use plenty of silicone rubber caulking to seal collector cover sheets.

Performance

Efficiency is about 30% to 35%.

Kansas Solid Block Wall

Developed at Kansas State University
by C.K. Spillman

This collector is an example of preheating ventilating air in a one pass system with storage, Fig 24.

The heart of this system is a 16″ thick wall of solid concrete blocks along the south side of a livestock building. The wall absorbs and stores solar heat. Advantages of this system include:

- Small space requirement.
- The vertical wall is steeper than the optimum latitude + 15° collector tilt angle, but it is good for winter months and does not collect ice and snow. In addition, a vertical surface receives little undesirable solar energy during the summer.
- In winter, heat is stored in the wall during the day and released at night.
- In summer, outside air cools the block at night. During the day, ventilating air passes through the blocks and is cooled on its way into the building.
- Enough fossil fuel is saved so the system may pay for itself faster than other solar investments. Typical savings can range from 1 to 2 gal of LP gas per ft² of collector surface per year.
- Multiuse possibilities. The collector can provide heat for grain drying in the fall and preheat ventilating air for the livestock building in the winter.

Construction

The construction steps are:

1. Erect a 16″ thick wall of solid concrete blocks on a concrete footing about 6″ outside of the south wall of the building. All blocks are perpendicular to the wall. Mortar only the horizontal joints and leave ⅛″-³⁄₁₆″ vertical gaps between the blocks.
2. Spray the south surface of the wall with two coats of flat black paint.
3. Attach vertical 2x2s to the south side of the wall.
4. Fasten a layer of transparent cover material to the 2x2s. Leave a 4″ air gap at the bottom.
5. Attach another framework of 2x2s over the first, sandwiching the cover material between them.
6. Fasten another layer of cover material to the outer 2x2s with battens and wood screws. Leave a screened 2″ air inlet at the top of this cover.

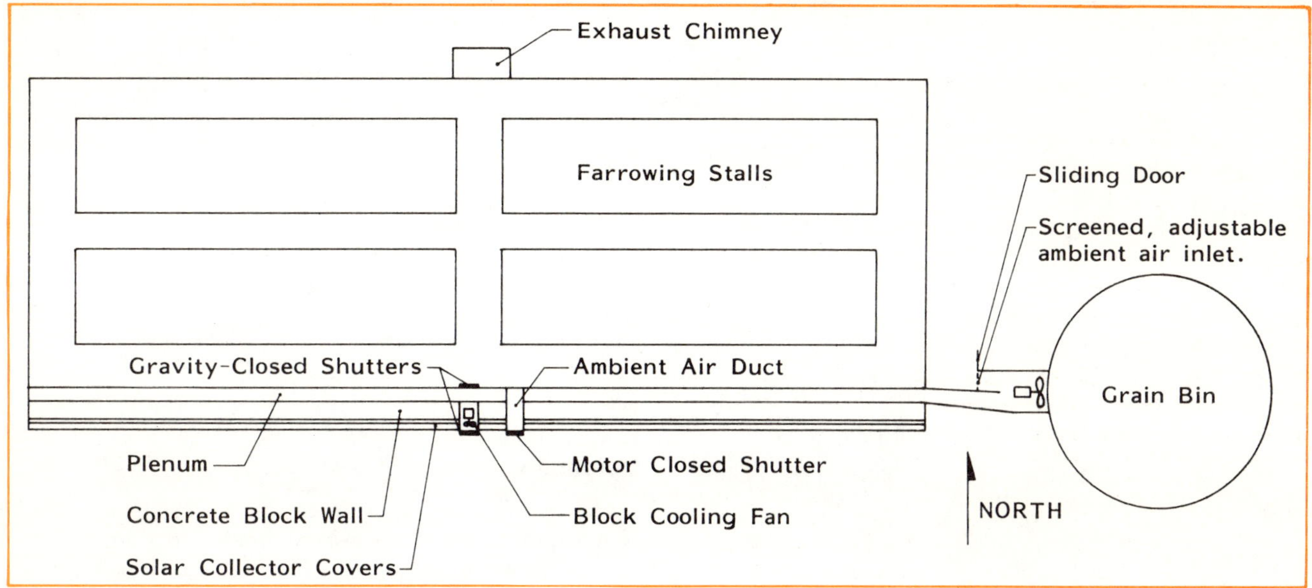

Fig 53. Kansas solid block wall—floor plan.

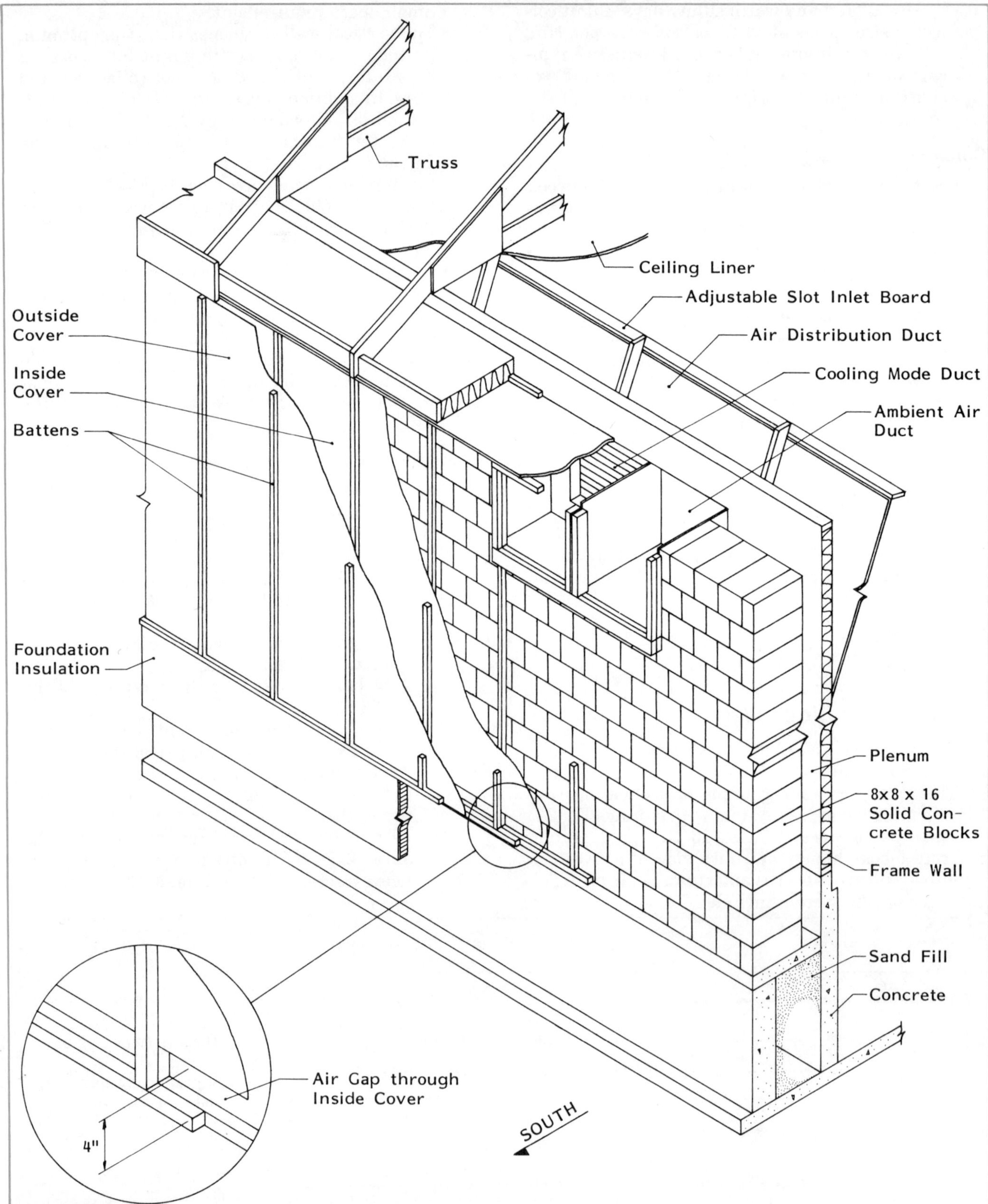

Fig 54. Kansas solid block wall—south wall detail.

If you decide to build this type of collector, order 20-Sow Solar Farrowing House, mwps-81902, from Midwest Plan Service. Construction details, more on how to use the system, and a wiring diagram for the controls are on the plan.

Operation

During winter, operate the collector/storage system for heating with stored heat. Outdoor air is pulled between the covers, through the blocks, and into the building. Heat is stored in the blocks during the day; ventilating air leaves them only slightly warmer than outdoor air. At night, ventilating air picks up heat from the warm blocks to help maintain building temperature.

On summer nights, operate the system for cooling. The building ventilating fan bypasses the collector wall, pulling cool air into the building through the outdoor air duct. A separate fan pulls cool night air through the blocks and exhausts it back outdoors. During the day operate the system as for heating—the blocks cool the ventilating air.

For grain drying, the crop drying fan pulls outdoor air through the collector, the plenum, and the duct to the grain bin. Ventilating air enters the building through the outdoor air duct to provide fresh air for the livestock building. Design the collector for its main purpose. If the pressure drop is greater than ¼″ water when all the grain drying air passes through the collector, open the slide to admit some outdoor air to the crop drying fan. See Fig 56.

System Performance

As tested at Kansas State, the block wall provided 32 Btu of heat storage per degree Fahrenheit temperature change per ft^2 of collector surface area. At 1 cfm per ft^2 of collector area, peak solar insolation was about 9 hr ahead of the peak temperature of the ventilating air and even further ahead of the peak temperature rise. Under Kansas conditions, average all-day collector efficiency was:

Airflow	Efficiency
1 cfm/ft^2	45%
2	55%
3	60%
4	62%
5	63%

About 15 to 20 ft^2 of collector area is needed per farrowing stall in the North Central Region.

Iowa Hollow Block Wall

Developed at Iowa State University
Description by W.F. Wilcke

This is another example of a one pass system with solar heat storage.

Description

This solar collector is a modification of KSU's solid concrete block wall. At the heart of the ISU collector is a conventional hollow concrete block wall on 4x4x8 concrete bricks spaced 8″ o.c. along the south side of a livestock building. Space the bricks to support block joints and ribs and to leave block cores open for airflow. In new construction, the blocks are usually a load-bearing wall. The inside surface is insulated and lined with animal-resistant material. For retrofit, set the wall on a footing cast along the south side of the building.

The block wall is both solar absorber and heat storage. Vertical wood furring strips are anchored to the south side of the blocks. Paint furring strips and block flat black (two coats) and screw clear greenhouse grade FRP to the strips for the collector cover. Install a screened air inlet (½″ hardware cloth) under the eave or at the top of the wall.

Operation

Exhaust fans, not in the south building wall, move air through the collector. Outside air enters the screened inlet, moves down the face of the blocks, up through their cores, and then through the building inlet system. On sunny days, air scrubs solar heat from the block face, but then loses part of it to the relatively cool block cores. During late afternoon, as outside temperatures fall, cold ventilating air is warmed by the stored heat in the block cores. Solar heat entry into the building is therefore delayed from mid-day until late afternoon and early evening.

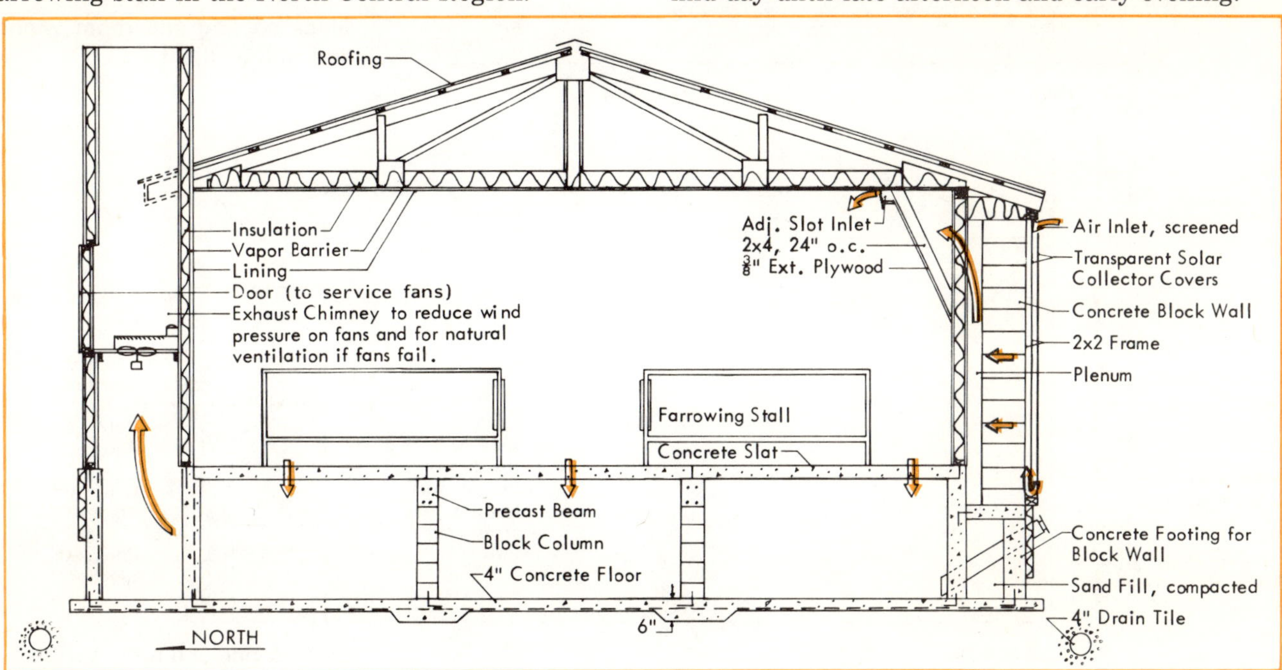

Fig 55. Kansas solid block wall—cross section.

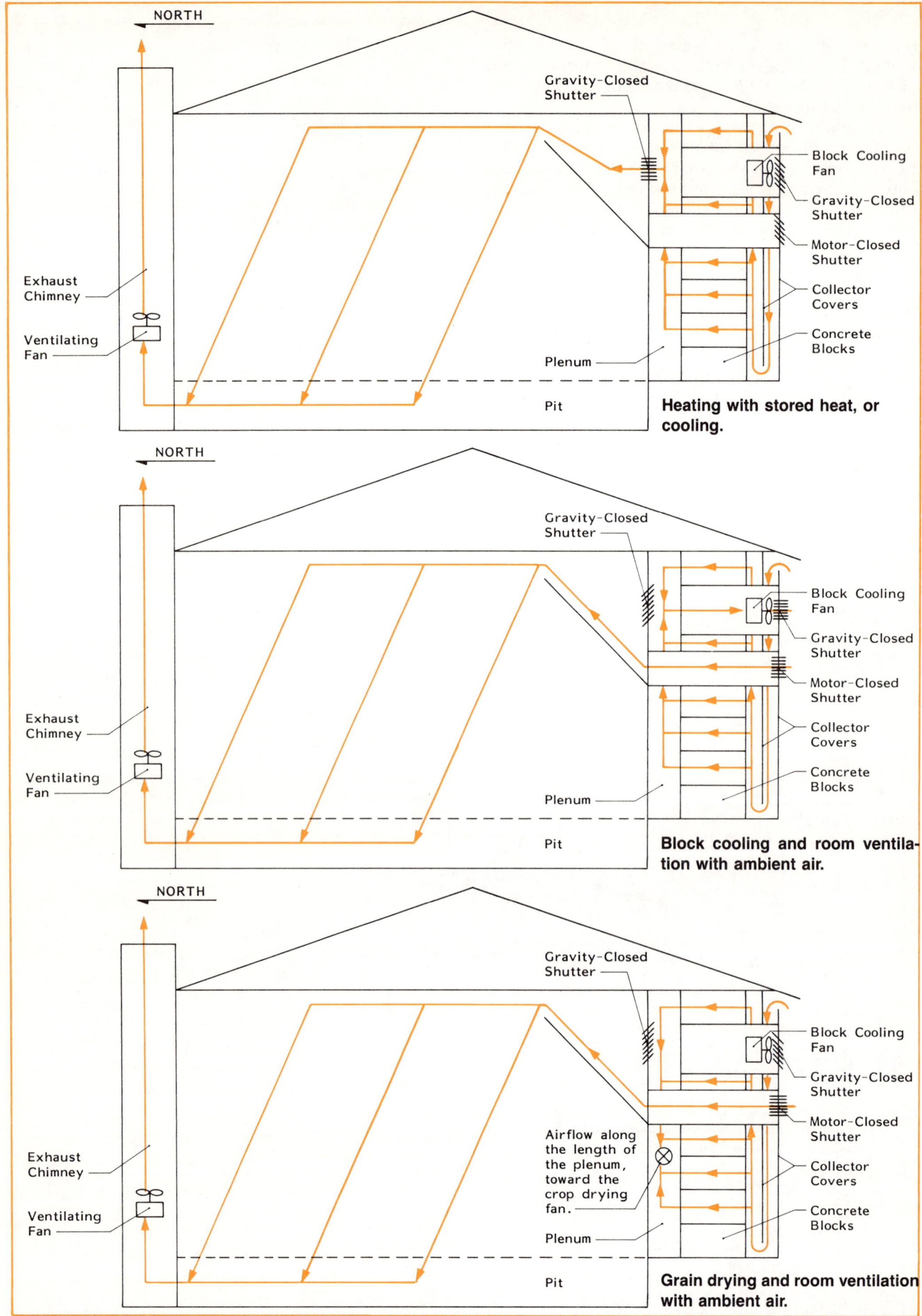

Fig 56. Kansas solid block wall—operation diagrams.

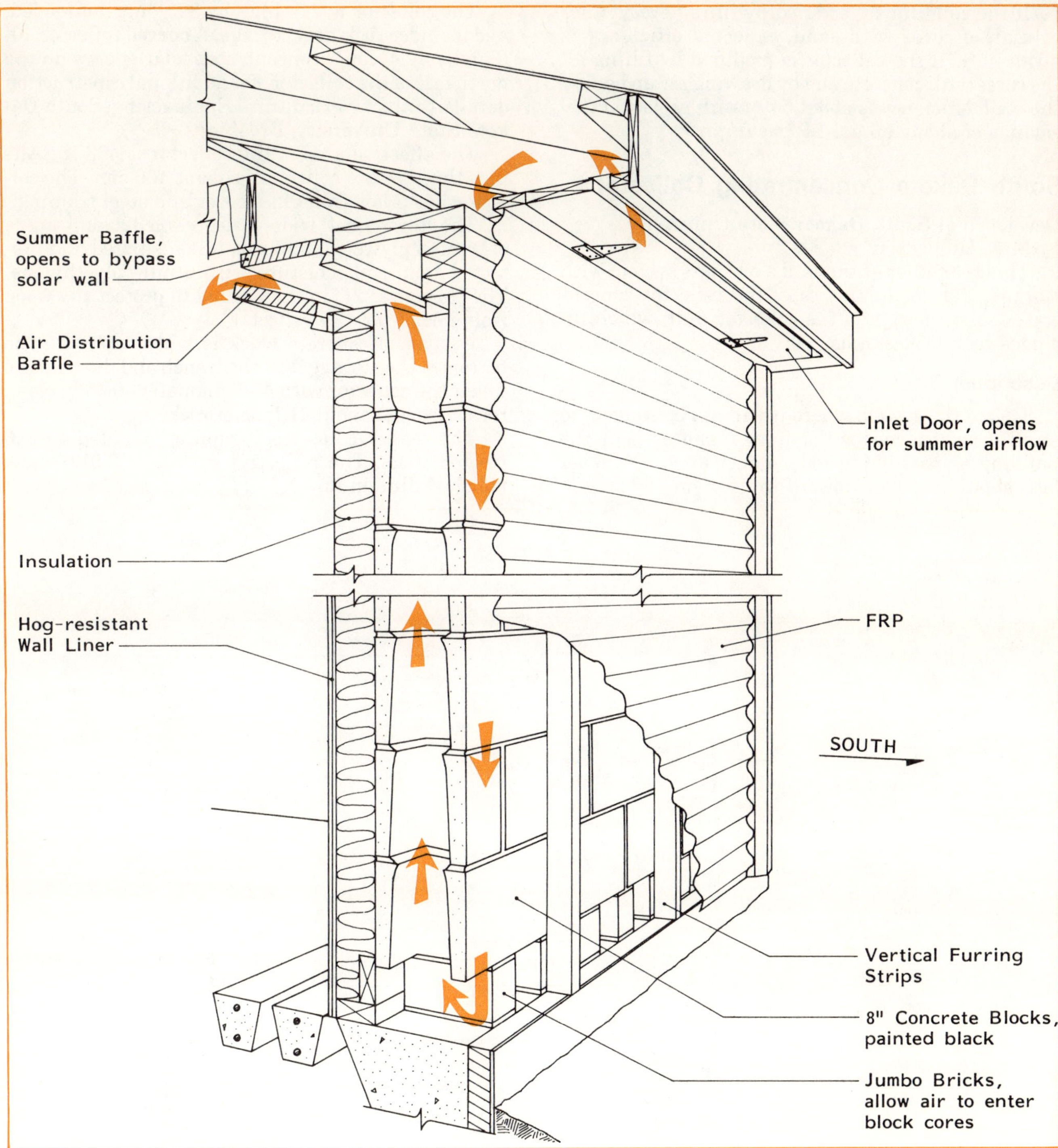

Fig 57. Iowa hollow block wall.

On warm days, the solar wall can cause building overheating, which activates large temperature control fans. To reduce wasting solar heat and electricity for the fans, provide a separate air inlet (opened manually or automatically) that bypasses the collector. In summer, bypass the collector and shade, cover, or vent it.

Variations

To increase the heat storage capacity of the block wall, increase the wall's mass by filling alternate cores with concrete, sand, or gravel. For sand, seal the bottom block so the sand does not run out.

For retrofit, you can set the block wall and foundation bricks 4″ away from the building and fill all block cores with concrete or mortar. Now, air moves down the south face of the blocks, under the wall, and up behind the wall through the 4″ plenum. Storage mass and collection efficiency are increased. Make sure the ends and top of the plenum are sealed and that the wall is adequately tied to the building.

Performance

The efficiency of the collector as illustrated is about 40%. If used for October to March heating, it can replace about 1 gal LP gas/ft²-yr.

If the collector is modified by filling every other column of cores with sand, expected efficiency is about 45%. If the collector is modified by filling all the cores with concrete and by drawing air up behind the wall, efficiency is about 55%, with projected fuel savings of about 1.5 gal LP gas/ft²-yr.

South Dakota Concentrating Collector

Developed at South Dakota State University
by M.A. Hellickson

This is another example of a one pass system with storage, Fig 25. In this case, the reflector concentrates solar energy on the collector unit, which includes rock heat storage.

Description

Locate the system where it will not be shaded, to avoid normal farmstead activities, and as near the building as possible to reduce duct losses. Exhaust fans should not blow toward the system.

The collector is 2.5' high by 32' long, dual-sided, and includes rock heat storage. A curved reflector, 10' high by 42.4' long, concentrates solar energy on the north face of the collector. For additional construction details, contact Agricultural Engineering, South Dakota State University, Brookings.

The effective area of the collector is 408 ft². Airflow through the collector is about 400 cfm. The collector has a low-iron glass cover and steel framing.

The storage is 4' wide, 2' deep, and 32' long. Made of treated plywood and 2x4s, it is insulated and protected from soil moisture with 6 mil polyethylene. The storage is 21" in the ground to protect the wood from heat from the reflector.

A layer of concrete block is laid 1" apart in the bottom of the storage. Fill the trench and the triangle under the collector with 4"-6" diameter smooth, clean rock. It holds about 11 tons of rock.

The reflector has steel framing mounted on 6x6 treated posts. The reflective surface is 0.012" thick polished aluminum.

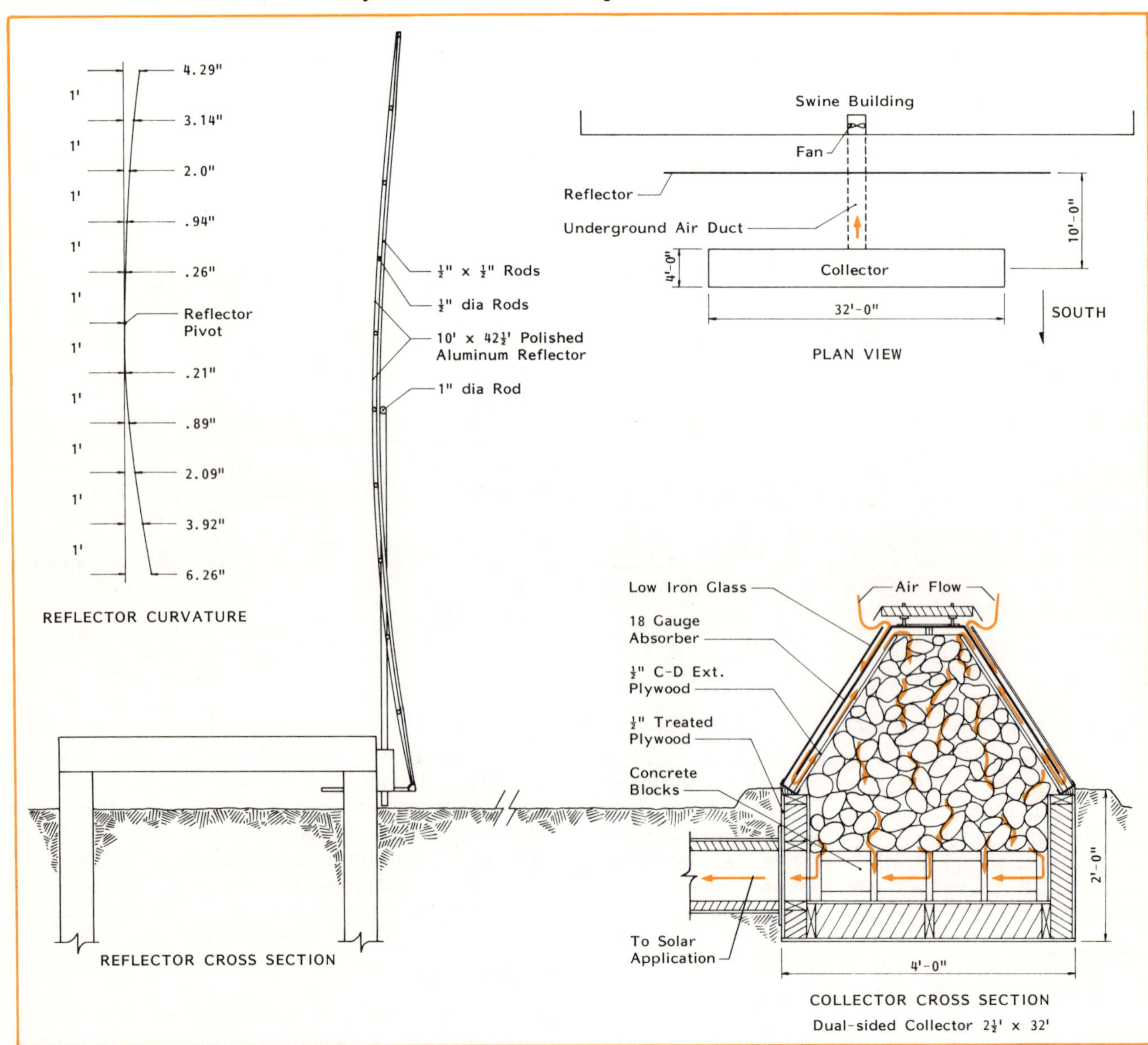

Fig 58. South Dakota reflector—plan and sections.

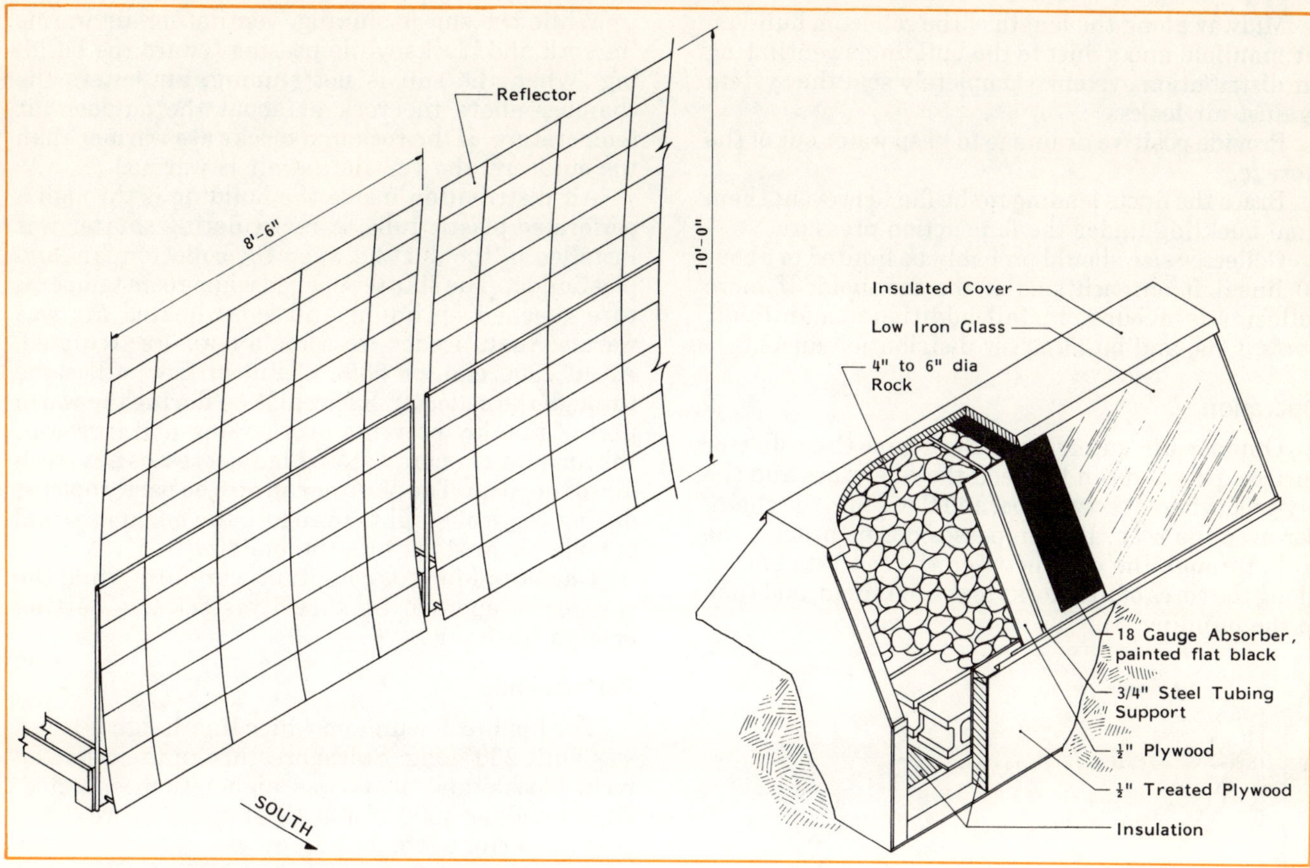

Fig 59. South Dakota reflector—pictorial views.

Size ducts for about 500 fpm air velocity for heating buildings. Build them of ½" treated plywood and rigid foam insulation. Glued and nailed construction helps maintain rigidity and therefore air tightness.

Operation and Maintenance

Adjust the reflector tilt to focus solar energy near the top of the glass at 9:00 a.m. and 3:00 p.m. and near the bottom of the glass at noon. Check reflector focus weekly. Adjust reflector tilt with the bolts fastening it to the support posts.

Clean the reflector with water and a sponge. Do not scrape the reflective surface.

Clean collector glass with mild soap and water or with a good glass cleaner. Brush snow from the glass as needed.

Remove anything that shades the system. Keep all flammable materials (wood, grass, straw, cloth, etc.) from the path of the reflected energy.

Efficiency

Average system efficiency is about 25%-30%. Test results with the size of unit shown were about 15 F temperature rise on the average with 400 cfm airflow through the collector.

Missouri Turkey Unit (Single Pass With Storage)

Developed at the University of Missouri-Columbia by R.E. Phillips

This is another one pass collector and storage system, Fig 26. It can be part of a new building or added to an existing one.

Description

The collector is supported on 2x4x8' studs 4' o.c. on a ³⁄₂ slope (56.3°). The absorber is ⅜" plywood painted black on the outside face. Install a 2x2 spacer above each 2x4 and midway between each pair of 2x4s. Fasten collector grade fiberglass on the spacers, leaving an air inlet slot near the top. Mount ribs in the plastic up and down the slope to shed snow. Air leaves the collector through 1"x10" slots in the absorber plate between the 2x2 spacers near the bottom of the absorber and enters the chamber above the gravel.

Support the collector on a foundation wall along its lower end and by the building at its upper end. Posts support preservative-treated tongue-and-groove (T&G) planking and 3" of extruded polystyrene insulation. Attachment to the building depends on building framing, but it must be weather resistant and airtight.

The solar heat storage unit has insulation next to the ground. Concrete blocks and 1½" diameter rock about 2' deep (about 1 ft³ or about 100 lb rock/ft² collector) from the storage. The blocks are part of the mass for energy storage and also form a duct for the ventilating air. Lay them without mortar and spaced ⅛"-¼" apart.

Midway along the length of the collector, build an air manifold and a duct to the building's ventilating air distribution system. Completely seal the system against air leaks.

Provide positive drainage to keep water out of the storage.

Brace the ducts leading to the fan to prevent them from buckling under the fan suction pressure.

Collector size should probably be limited to about 60 lineal ft on each side of the manifold. If more collector is needed, install additional manifolds, ducts, fans, and building air distribution ducts.

Operation

Outdoor air enters near the top of the collector surface, passes down between the fiberglass and the plywood where it is warmed, and flows into the chamber over the rock. It then passes down through the rock, through the gaps between the concrete blocks, along the cores of the blocks to the manifold, and then to the building.

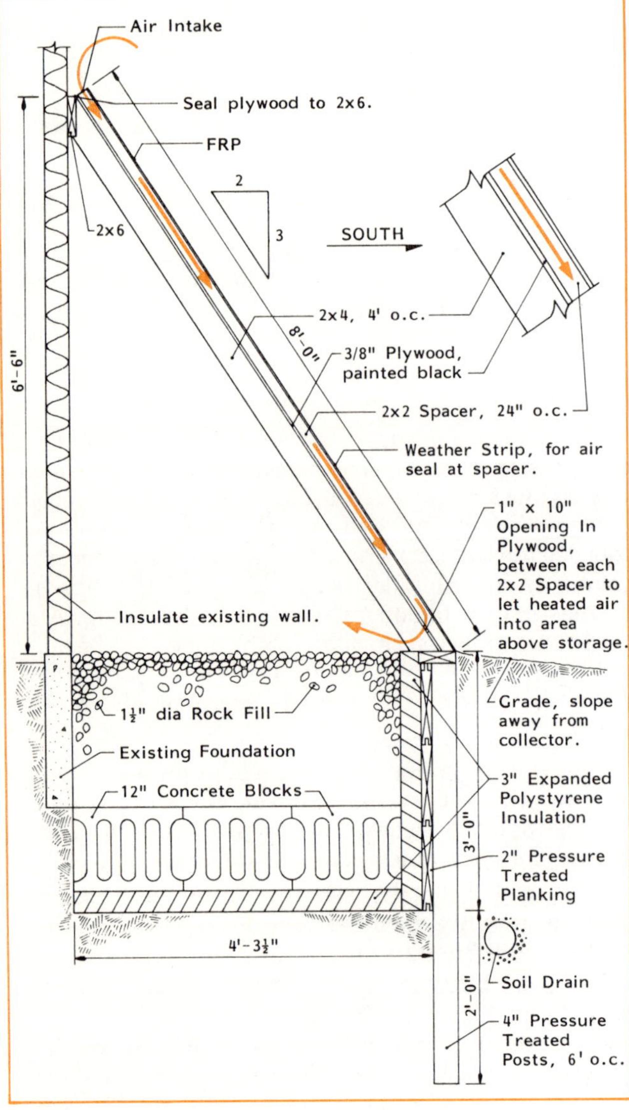

Fig 60. Missouri collector.

While the sun is shining, ventilating air warms the rock and blocks while passing toward the building. When the sun is not shining, air enters the chamber above the rock at about the outdoor air temperature. If the rock and blocks are warmer than the outdoors, the ventilating air is warmed.

Air distribution inside the building is through a perforated plastic tube. A recirculating shutter was installed in the duct between the collector, fan, and plastic duct. The shutter opened when room temperature reached optimum and solar-heated air was warmer than needed, so some air was recirculated. About 1900 cfm, or 60%, of full airflow still came through the collector. Recirculation during the warm part of the day prevents overheating and increases the amount of energy stored but decreases net fresh air to the room. The shutter can also be partly opened during very cold nights to reduce the amount of cold outside air brought into the building.

Caution: Monitor conditions carefully while the shutter is open to be sure livestock are getting enough fresh air.

Performance

The field-tested unit on which this design is based was built 250' long. Static pressure in the collector, rock, blocks, and ducts was higher than expected. Fans rated for 4000 cfm at ⅛" static pressure delivered 3200 cfm to the building.

Airflow was 1.6 cfm/ft^2 of collector.
• Air speed was about:
110 fpm through the collector.
180 fpm through the 1"x10" openings.
845 fpm through the blocks at the manifold.

For the two test winters in Missouri (latitude 37°), the average temperature rise from outdoor air to fan was about 20 F. The storage rock was warmed as much as 30 F above outside temperature.

The system efficiency for the unit (from collector to fans) was about 60%.

The system saved about 1 gal propane per day for each 100 ft^2 of collector surface during December through January.

Illinois Lean-to Collector with Heat Storage

Developed at the University of Illinois
by W.H. Peterson

This collector is an example of a system with a collect/store loop and one storage/building pass, Fig 26. It has a lean-to built over a rock pile along the south side of a building. Solar energy passing through the greenhouse grade fiberglass cover is absorbed by the building wall and rocks. A charging fan circulates sun-heated air down through the rock pile from the space above as long as the air is warmer than the rocks at the bottom of the pile. The winter ventilating fan continuously pulls outside air through a 1½" slot in the cover, through the rock pile, and into the building.

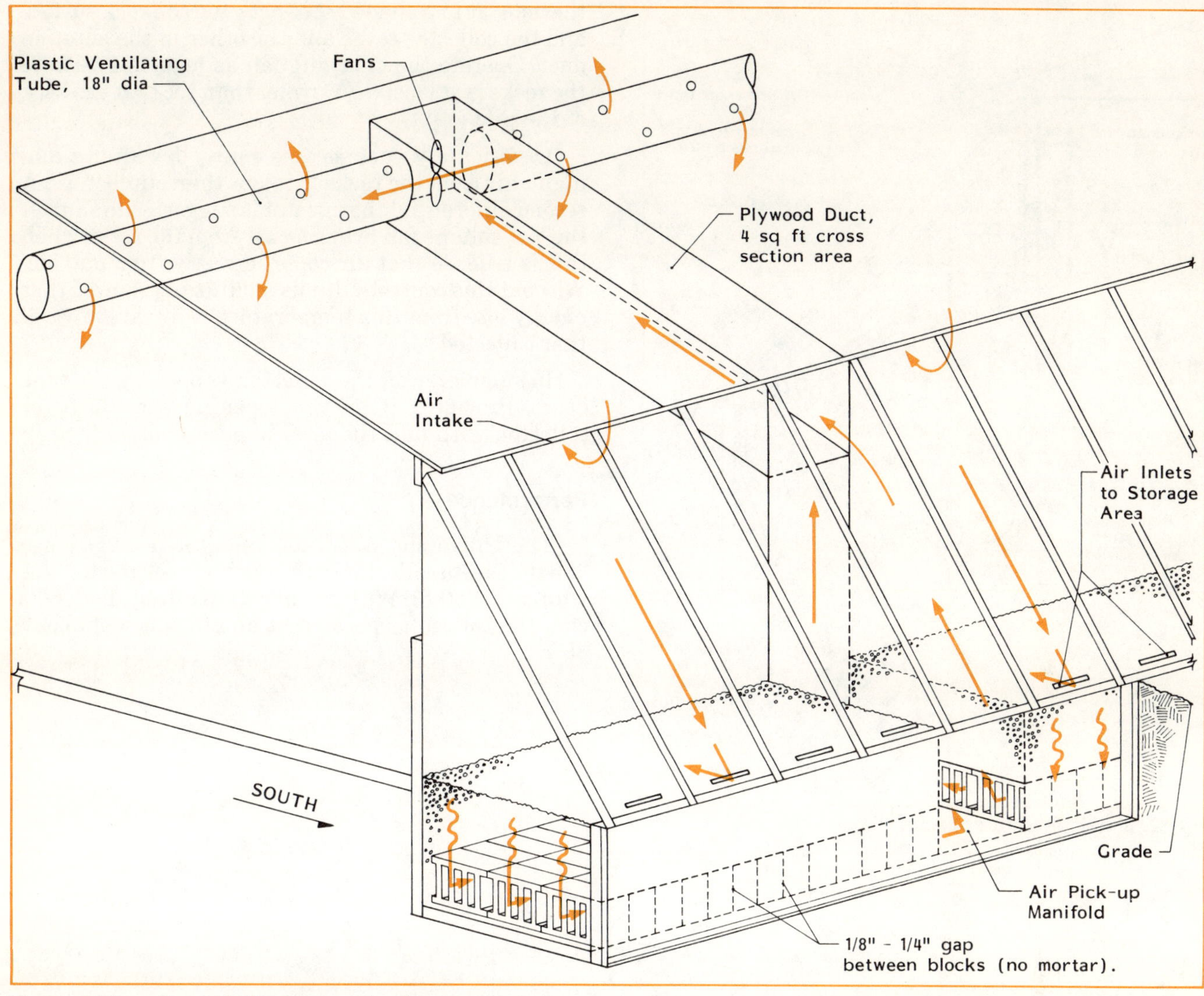

Fig 61. Airflow through the Missouri collector.

Design and Construction

Collector width is 8′; collector height depends on building height. Attach the collector rafters below the building eave. Enclose as much of the south wall as practical. Paint it black to maximize solar absorption. If the sidewall is higher than 10′, you may not need a collector all the way to the eave. To maintain adequate collector tilt angle, reduce the base width to less than 8′ if the sidewall is less than 7′ high. A 7′ high, 8′ wide collector has a cover tilt angle of 40°— about the minimum for winter solar energy collection and snow removal in northern regions.

Calculate required collector length by dividing required collector area (see Design section) by the length of the fiberglass cover sheets. Because construction costs are much higher when rock pile collectors are built separate from a building, build this collector type only along existing structures. Limit collector length to the length of the supporting building. For a collector longer than 50′, build it in two shorter units to avoid air distribution problems in the rock pile.

For heat storage, select smooth, graded rocks about 4″ in diameter. You will need about 1 to 1.5 ft³ (100 to 150 lb) of rock per ft² of collector area. Shape the pile so air travel distance from the surface of the rocks to the aeration duct is about uniform. Spray the surface of the pile and the building sidewall with two coats of flat black paint.

Because rocks restrict airflow, size both fans to deliver necessary airflow against ¼″ water pressure instead of the usual ⅛″. Size the building air fan for the minimum winter ventilation needs of the housed animals. Equip this fan with a gravity back-draft shutter and a thermostat that stops the fan when the building gets too warm. Provide other fans and a thermostatically controlled air inlet to ventilate the building in warm weather.

Size the charging fan for three times as much air (at ¼″ water pressure) as the ventilating fan; the bearing must permit vertical mounting. Equip the charging fan with a gravity back-draft shutter and a plywood baffle to divert air toward the ends of the collector. Place one sensing element of a differential

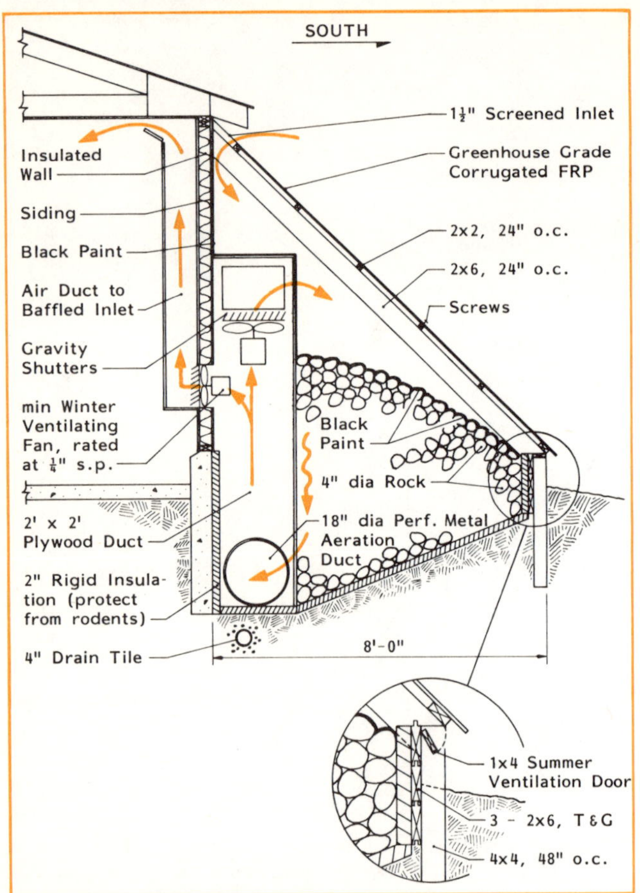

Fig 62. Illinois lean-to collector.

thermostat in a shaded spot between the rock surface and the collector cover and the other in the aeration duct. Operate the charging fan as long as air above the rocks is at least 5 F warmer than rocks at the back of the pile.

Occasionally, such as on a sunny day after a cold night, air from the rocks is colder than outdoor air. A second differential thermostat is suggested to shut off the low-volume fan bringing air from the rocks when this is true, so that air comes directly from outside. Without this control, Illinois monitoring showed that energy loss from this temperature reversal is 26% of that collected.

In summer, vent the collector by opening the vent doors. Provide 1 ft² of vent opening per 100 ft² of collector, with half the area in each end.

Performance

The University of Illinois monitored a 28′ long, 336 ft² lean-to collector with 780 ft³ of rock during the winter of 1980-81. With a ventilation rate of about 600 cfm, the collector operated at an efficiency of about 35%.

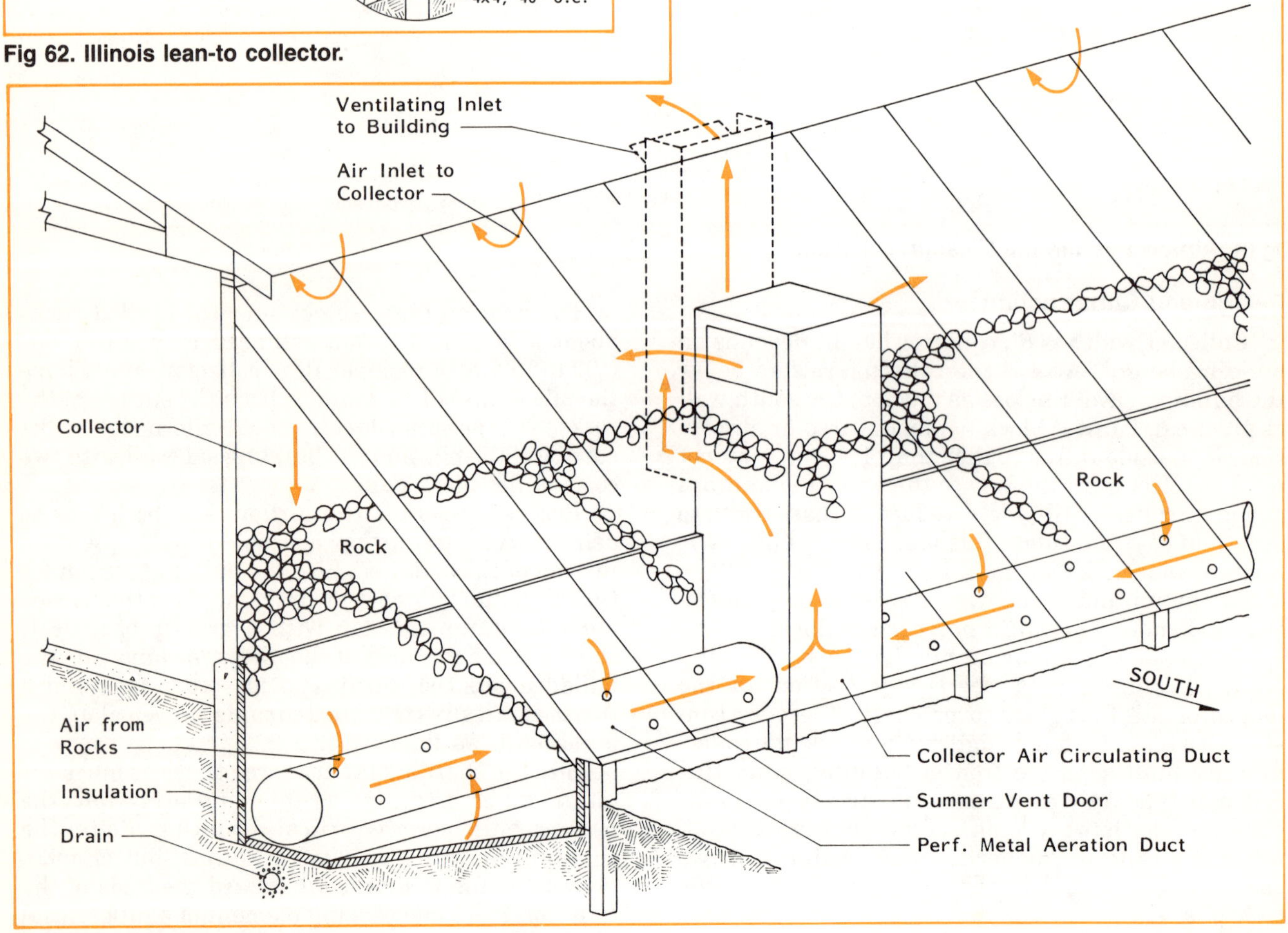

Fig 63. Airflow diagram for Illinois lean-to collector.

Nebraska Modified Open-Front Swine Nursery

Developed at the University of Nebraska
by G.R. Bodman

This building is an example of both passive solar (Fig 27) and indirect heating of a livestock unit (Fig 29). Solar energy provides all supplemental heat during normal winters. Small pigs may need heat lamps after two weeks of cold, overcast weather.

Passive Solar System

Windows along the entire south side of the building provide passive solar heating and drying during the day. Some direct solar energy is absorbed and stored in concrete floors and partitions and is released at night. Night heat loss is reduced with double-layer windows. The windows are made of sun-resistant materials and are easily adjusted.

The building has no mechanical ventilation or backup heater. Energy use is minimal, but careful management is needed to keep the ventilating panels adjusted for optimum indoor temperatures and moisture levels.

The south roof overhang shades the solar windows in summer to prevent overheating. See collector shading in the Fundamentals section for overhang width for your latitude. Open the lower windows in the south wall for additional natural summer ventilation.

For construction details, order plan NE 10.726-37 from Agricultural Engineering Plan Service, 215 Chase Hall, University of Nebraska, Lincoln, NE 68583. The plan is for a 23′x116′, 550-head, solar swine nursery, and includes adapting it to 100- to 600-head capacity.

Active Solar System

The building has an active solar system for floor heating. A fan circulates solar-heated air in a closed loop between a collector set along the south side of the building and a heat storage bed under the floor along the north side of the building.

Collector

The single-channel, covered plate collector has a wood frame, plywood backplate, and two Tedlar covers. Dark-colored, prepainted metal siding or roofing is recommended for the insulated absorber. The collector is 40″ wide (36″ wide absorber) and the full length of the building. It is at a 60° tilt angle on a concrete slab.

Design air velocity through the collector is 1000 fpm and design airflow is 2 cfm/ft^2. For a 36″ wide absorber, required air channel depth inside the collector is:

Channel depth (in.) = 0.024 x Collector length (ft)

Rip the 2″ lumber side supports to give the proper channel depth if necessary.

Build and seal the collector carefully to prevent air leaks, which greatly reduce the efficiency of closed loop systems.

Storage

Solar energy is stored by forcing solar-heated air from the collector through the cores of masonry units under a bed of sand, Figs 64 and 65. The heat warms the blocks and passes up through the sand to warm the floor. Insulate around the storage.

Because air circulates only when solar energy is available and always in the same direction, the floor is usually warmer at the fan, or pressure, end of the storage. Keep youngest animals closest to the fan and move them along the building as they grow.

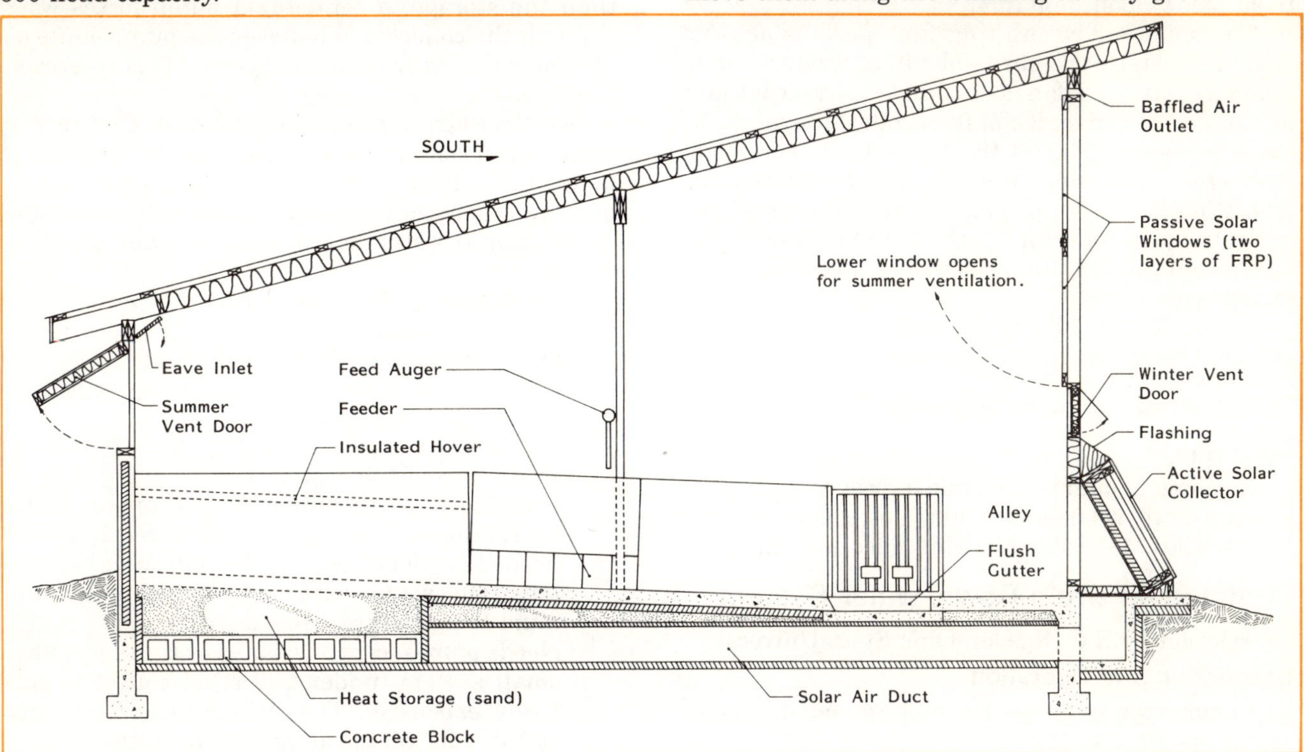

Fig 64. Nebraska solar MOF swine nursery.

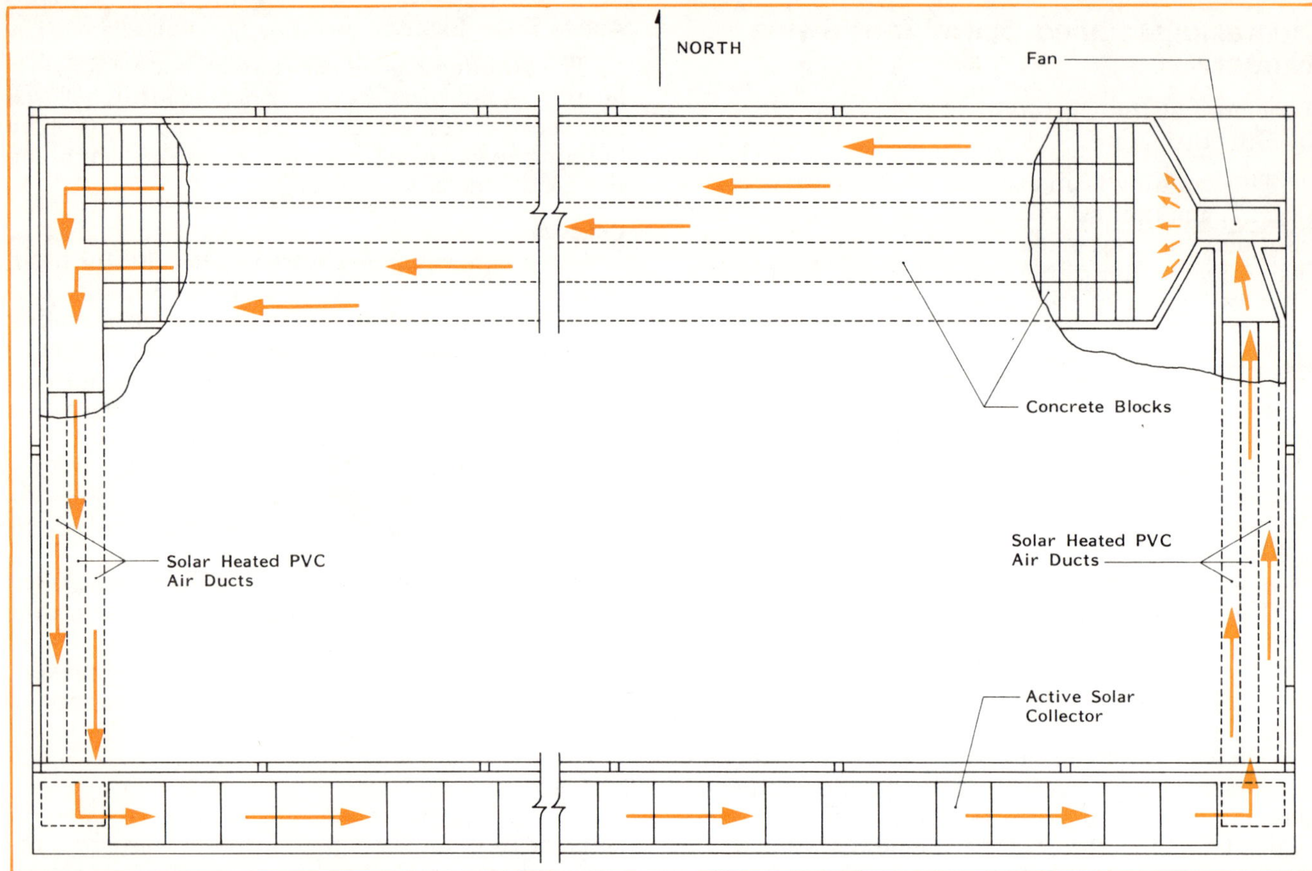

Fig 65. Airflow in Nebraska solar MOF nursery.

Air movement

The circulation fan is usually in the duct in the northeast corner of the floor. Ducts are PVC tubes wrapped in fiberglass insulation. Install enough tubes of proper size to keep average air velocity in them below about 600 fpm.

The solar air circulating fan operates against fairly high static pressure—about ¾″ water column. Select a centrifugal fan (also called squirrel cage fans or blowers) to deliver 2 cfm/ft² of collector against ¾″ static pressure. Control the fan with a thermostat with a remote sensing bulb placed in the shade in the transition 1″-2″ from the end of the absorber at the outlet from the collector. Set the thermostat so the fan runs when air from the collector is above 95 F.

Performance

About 25% to 30% of the radiation striking the collector heats the in-floor storage.

Direct Heating from Storage

Solar Shop

A complete plan is available for a shop with space heating with solar-heated air. Get a copy of plan mwps-81901, 48′ Solar Machine Shed and Shop.

Commercial Drain-Down Water System

Description by R.C. Reeder, Ohio State University

Description and Operation

Commercial solar systems of the automatic drain-down type are available.

The collectors are usually built into the south-facing roof during construction. Heat storage is in an insulated tank, often under the floor of a service room.

When the collector is several degrees warmer than the storage, a centrifugal pump forces water through the collector. Whenever the pump shuts off, all water drains from the collector pipes to prevent freezing.

For floor heating, such as in farrowing or nursery buildings, a separate pump circulates water through pipes in the floor. If stored water is cooler than required, automatic valves shunt the water through a conventional water heater and the storage is bypassed.

Advantages of the system are:
- Uses solar heat year-round.
- Occupies little building floor space.
- Has a reasonable initial cost for systems with 500 ft² or more of collector.
- Efficiently stores solar energy for later use.

Limitations of the system are:
- Has limited hours of winter solar energy collection because of the high temperatures (105-110 F) needed for floor heating. An improved heat distribution system can help by lowering the minimum temperature a few degrees.
- Needs pumps, controls, valves, etc., that make a small system (under 200 ft² of collector) relatively expensive. The system is perhaps most suitable for 500 to 1000 ft² collectors.

- Requires accurate installation—slopes of pipes for drain-down, watertightness, etc.
- Requires professional installation—not a do-it-yourself system.
- Needs maintenance for the several working parts.

Components and Construction Details

Have the unit installed according to the manufacturer's plans and specifications.

Collector panels have ½" copper pipes spaced 6" apart and covered with a thin copper sheet painted with a solar-selective coating. The panels are connected together on all or part of the south slope of the roof and plumbed to supply and return lines. Some prefabricated panels can be mounted on the ground. All lines must slope slightly to one end and drain to the storage.

FRP roofing or other suitable material covers the collectors.

Provide storage capacity of 2 to 3 gal/ft² collector area.

The storage is usally a concrete tank installed under a floor. Insulate tank top, bottom, and sides to at least R = 12 with extruded polystyrene insulation.

Because water tends to stratify in the storage according to temperature, collector loop water is taken from the bottom of the tank and returned to the top of the tank. Floor heating water goes from the top of the tank to the floor loops and returns to the bottom of the tank.

The return line from the collector is above the storage tank water level to permit lines to drain even if the drain-down valve fails to open.

If floor heating pipes are spaced closer together (8"-10" apart), slightly lower storage temperatures will maintain adequate floor temperature. The lower operating water temperature improves collector system performance.

System Performance

Tests in Illinois, Missouri, and Ohio indicate yearly energy savings of about 1 gal LP gas/ft² of collector for farrowing or nursery floor heating systems.

Performance is closely related to hours of solar collection. Year-round use is needed to pay for this type of system. In Ohio, heat is being stored 1½ to 2 hr/day in winter, 3 to 4 hr/day in spring and fall, and about 5 hr/day in summer.

On bright days between 10 a.m. and 3 p.m., daily collector efficiency is about 30%-35%. On cloudy days, even though there is diffuse solar energy available, the system runs little, if at all. Year-round efficiency is about 20%. Efficiency drops as the storage temperature approaches 100 F above outdoor temperature. Winter storage temperature is usually less than 120 F, because the heat is used almost as rapidly as it is collected. Summer water temperatures can reach 165 F.

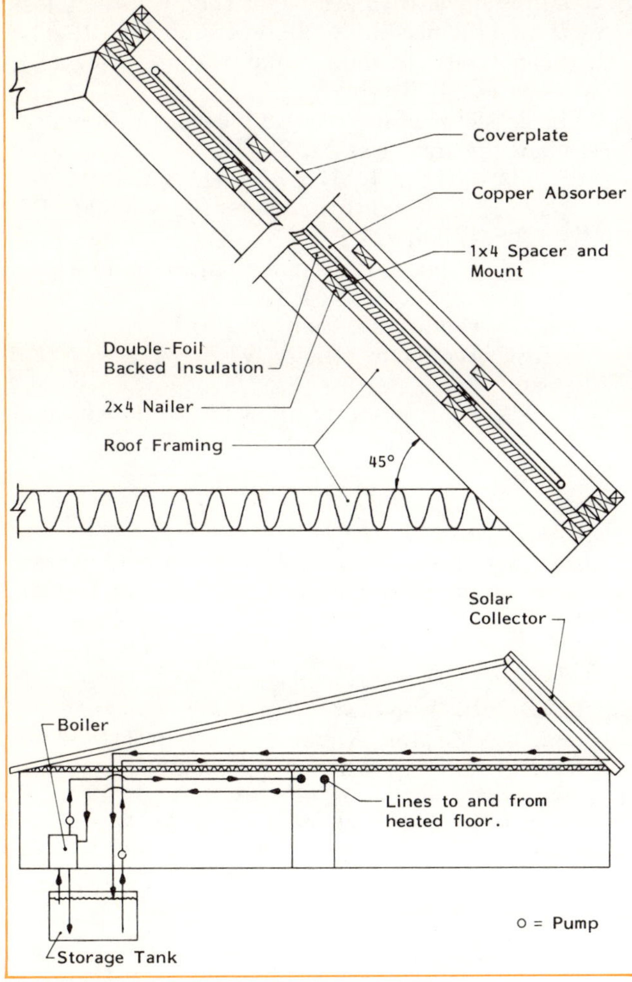

Fig 66. Commercial drain-down water system.
Roof-mounted commercial solar collector.

APPENDIX

Data are for average clear days—on exceptionally clear days, values 15% greater are possible. Values include direct plus diffuse radiation, Btu/hr-ft², on south-facing surfaces. This table was developed by MWPS with technique from *1978 ASHRAE Applications Handbook*. Measure collector tilt angles from horizontal. Surfaces with a latitude + 15° tilt angle receive the most energy during the space heating season. Use this table to calculate sunny day collector output at your latitude. Table times are solar time.

Example:

Find the maximum temperature rise from a 2400 ft² solar collector on a south-facing ⁴/₁₂ roof slope in January near Bismarck, ND. Airflow is 2,500 cfm.

Answer:

The maximum temperature rise is at noon on sunny days. Bismarck's latitude is a little less than 47°—use Table 26 for 46° on Jan. 21. The noon hour, clear day solar radiation on a ⁴/₁₂ slope is 201 Btu/hr-ft².

The collector plan gives an average all-day efficiency of 15%. Use Eq 7.

$$TD = EFF \times RAD \times A \div (1.1 \times Ccfm)$$
$$= 0.15 \times 201 \times 2400 \div (1.1 \times 2,500)$$
$$= \text{about 26 F}$$

TD = maximum temperature rise, F
A = collector area, ft²
RAD = solar radiation, Btu/hr-ft²
EFF = collector efficiency, %
Ccfm = collector airflow, cfm
1.1 = a constant to allow for air density and specific heat of air, Btu/F-hr-cfm.

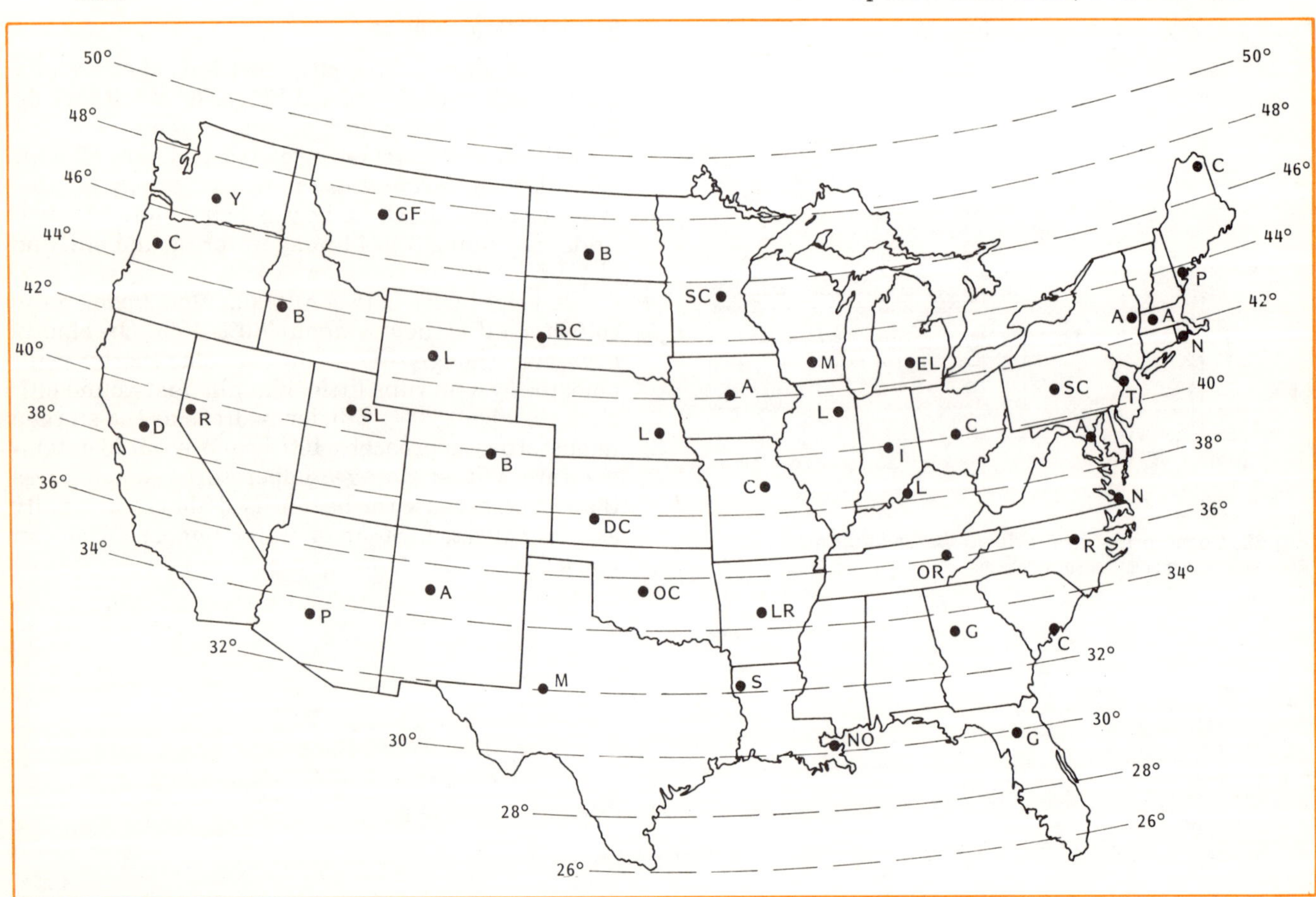

Fig 67. Latitudes and data locations for Tables 26 and 27.
Solar radiation data for the cities shown are in Table 27.

Table 26. Hourly, clear day solar radiation.

26° North Latitude

Date	AM	PM	0°	4/12	Lat.	Lat. + 15°	90°
Jan 21	7	5	6	12	14	18	21
	8	4	76	110	121	137	125
	9	3	143	190	204	221	177
	10	2	195	250	265	282	211
	11	1	228	287	303	320	231
	Noon		239	300	316	333	237
Daily total Btu/day-ft²			1540	2002	2135	2294	1771
Feb 21	7	5	29	41	44	49	44
	8	4	108	134	141	148	105
	9	3	178	213	222	227	147
	10	2	232	273	283	287	177
	11	1	266	311	321	325	195
	Noon		277	324	334	337	201
Daily total Btu/day-ft²			1907	2273	2360	2414	1542
Mar 21	7	5	58	62	62	59	28
	8	4	138	149	150	144	67
	9	3	207	226	227	219	101
	10	2	261	285	287	276	127
	11	1	295	322	324	312	144
	Noon		306	335	337	325	149
Daily total Btu/day-ft²			2226	2427	2441	2350	1087
Apr 21	6	6	8	5	4	4	2
	7	5	84	75	70	56	9
	8	4	160	155	148	129	19
	9	3	226	225	219	195	46
	10	2	276	280	273	247	68
	11	1	308	314	308	280	82
	Noon		319	326	320	292	87
Daily total Btu/day-ft²			2448	2440	2369	2119	545
May 21	6	6	26	15	10	10	5
	7	5	100	83	73	51	12
	8	4	172	156	144	116	14
	9	3	233	220	209	176	16
	10	2	281	271	259	224	29
	11	1	310	303	291	254	41
	Noon		320	314	302	265	45
Daily total Btu/day-ft²			2570	2414	2282	1933	285
Jun 21	6	6	32	19	13	12	7
	7	5	105	84	73	49	13
	8	4	175	154	142	111	16
	9	3	234	217	204	168	17
	10	2	280	265	252	214	18
	11	1	308	296	283	243	27
	Noon		318	306	293	253	31
Daily total Btu/day-ft²			2591	2382	2231	1850	230

Date	AM	PM	0°	4/12	Lat.	Lat. + 15°	90°
Jul 21	6	6	25	15	11	10	6
	7	5	99	82	72	51	13
	8	4	170	154	143	115	16
	9	3	230	217	206	174	17
	10	2	277	267	255	220	29
	11	1	306	298	287	250	41
	Noon		316	309	297	260	45
Daily total Btu/day-ft²			2535	2380	2250	1906	293
Aug 21	6	6	8	5	4	4	2
	7	5	82	74	68	55	11
	8	4	156	151	145	126	20
	9	3	221	220	213	190	46
	10	2	270	273	266	241	67
	11	1	301	306	300	273	81
	Noon		311	318	312	284	85
Daily total Btu/day-ft²			2390	2380	2310	2066	544
Sep 21	7	5	55	58	58	56	27
	8	4	133	143	144	138	65
	9	3	200	218	219	210	98
	10	2	253	275	277	266	123
	11	1	285	311	313	301	139
	Noon		296	324	325	313	145
Daily total Btu/day-ft²			2151	2340	2351	2260	1052
Oct 21	7	5	27	37	40	45	39
	8	4	104	129	135	141	101
	9	3	173	206	214	220	142
	10	2	226	266	275	279	172
	11	1	259	303	312	316	190
	Noon		270	316	325	328	196
Daily total Btu/day-ft²			1853	2202	2284	2333	1489
Nov 21	7	5	5	11	13	16	19
	8	4	75	108	118	133	121
	9	3	141	187	200	217	174
	10	2	193	246	261	278	208
	11	1	225	283	299	315	227
	Noon		237	296	312	328	234
Daily total Btu/day-ft²			1521	1970	2101	2253	1736
Dec 21	7	5	0	2	2	3	4
	8	4	63	98	110	127	126
	9	3	128	178	193	215	185
	10	2	179	237	254	276	220
	11	1	211	274	292	313	241
	Noon		222	286	305	326	247
Daily total Btu/day-ft²			1388	1867	2013	2200	1802

Hourly clear day radiation, Btu/hr-ft²

Table 26. Hourly, clear day solar radiation, continued.

28° North Latitude

Date	AM	PM	0°	⁴/₁₂	Lat.	Lat. + 15°	90°

Hourly clear day radiation, Btu/hr-ft²

Date	AM	PM	0°	⁴/₁₂	Lat.	Lat.+15°	90°
Jan 21	7	5	3	6	8	10	12
	8	4	69	103	117	132	122
	9	3	135	183	201	218	179
	10	2	186	242	262	279	215
	11	1	218	279	300	317	236
	Noon		229	292	313	330	243
Daily total Btu/day-ft²			1457	1924	2093	2246	1775
Feb 21	7	5	26	37	41	46	42
	8	4	103	130	139	145	107
	9	3	171	208	220	225	151
	10	2	224	268	281	285	183
	11	1	258	306	319	323	202
	Noon		269	319	332	335	209
Daily total Btu/day-ft²			1838	2222	2336	2388	1584
Mar 21	7	5	56	61	61	58	29
	8	4	135	148	149	143	71
	9	3	203	224	226	217	107
	10	2	256	282	285	275	135
	11	1	289	319	323	311	153
	Noon		300	332	336	323	159
Daily total Btu/day-ft²			2181	2405	2429	2336	1152
Apr 21	6	6	10	7	5	4	2
	7	5	85	77	70	56	9
	8	4	160	156	148	128	24
	9	3	224	226	218	194	53
	10	2	274	280	273	246	76
	11	1	305	314	307	279	91
	Noon		315	326	319	291	97
Daily total Btu/day-ft²			2434	2450	2364	2113	615
May 21	6	6	29	18	11	10	6
	7	5	103	86	73	51	12
	8	4	173	158	145	116	15
	9	3	234	223	209	176	17
	10	2	280	273	259	223	37
	11	1	309	305	291	254	51
	Noon		319	316	302	264	55
Daily total Btu/day-ft²			2579	2445	2283	1930	335
Jun 21	6	6	36	22	14	13	7
	7	5	108	88	74	49	13
	8	4	176	158	142	111	16
	9	3	235	220	203	168	17
	10	2	279	268	252	213	24
	11	1	308	298	282	242	36
	Noon		317	309	293	252	41
Daily total Btu/day-ft²			2608	2421	2233	1849	273
Jul 21	6	6	29	18	12	11	6
	7	5	102	85	73	51	13
	8	4	171	156	143	115	16
	9	3	230	220	206	173	18
	10	2	276	269	255	220	37
	11	1	305	300	286	249	50
	Noon		314	311	297	260	55
Daily total Btu/day-ft²			2543	2411	2251	1905	341
Aug 21	6	6	9	6	5	4	2
	7	5	83	75	68	55	11
	8	4	156	152	144	125	25
	9	3	219	220	212	189	52
	10	2	267	273	266	240	75
	11	1	298	306	299	272	90
	Noon		308	318	311	283	95
Daily total Btu/day-ft²			2378	2390	2306	2060	611
Sep 21	7	5	53	57	57	55	28
	8	4	130	142	143	137	68
	9	3	196	216	218	209	103
	10	2	248	273	275	265	130
	11	1	280	309	312	300	148
	Noon		291	321	324	312	153
Daily total Btu/day-ft²			2108	2317	2337	2246	1111
Oct 21	7	5	24	33	37	41	37
	8	4	99	124	133	139	102
	9	3	167	202	212	218	146
	10	2	218	261	273	277	178
	11	1	251	298	310	314	197
	Noon		262	310	323	326	203
Daily total Btu/day-ft²			1785	2150	2258	2306	1527
Nov 21	7	5	2	6	7	9	11
	8	4	68	100	114	128	119
	9	3	133	180	197	214	175
	10	2	184	239	258	275	211
	11	1	216	276	296	312	232
	Noon		227	288	309	325	239
Daily total Btu/day-ft²			1438	1893	2058	2205	1739
Dec 21	7	5	0	0	0	0	0
	8	4	55	89	104	121	121
	9	3	119	170	189	210	185
	10	2	169	229	251	272	223
	11	1	201	265	289	310	245
	Noon		211	278	302	323	252
Daily total Btu/day-ft²			1303	1788	1972	2153	1801

Table 26. Hourly, clear day solar radiation, continued.

30° North Latitude

Date	AM	PM	0°	4/12	Lat.	Lat. + 15°	90°
			Hourly clear day radiation, Btu/hr-ft²				
Jan 21	7	5	1	2	3	4	4
	8	4	62	95	111	126	119
	9	3	127	175	197	214	180
	10	2	177	235	259	276	218
	11	1	208	271	298	314	240
	Noon		219	284	311	327	248
	Daily total Btu/day-ft²		1373	1845	2051	2197	1775
Feb 21	7	5	23	33	38	42	39
	8	4	98	125	136	143	108
	9	3	165	203	217	223	155
	10	2	217	263	279	283	189
	11	1	249	300	317	321	209
	Noon		260	313	330	333	216
	Daily total Btu/day-ft²		1767	2167	2310	2361	1621
Mar 21	7	5	55	60	60	57	30
	8	4	131	146	148	142	74
	9	3	199	221	225	216	113
	10	2	250	280	284	273	142
	11	1	283	316	322	310	161
	Noon		294	329	334	322	168
	Daily total Btu/day-ft²		2133	2379	2415	2323	1214
Apr 21	6	6	12	8	5	5	2
	7	5	86	78	70	56	9
	8	4	159	157	148	128	29
	9	3	222	226	217	194	60
	10	2	271	280	272	246	85
	11	1	301	314	306	279	101
	Noon		312	326	318	290	106
	Daily total Btu/day-ft²		2418	2457	2359	2107	683
May 21	6	6	32	20	12	11	6
	7	5	105	89	74	51	12
	8	4	174	161	145	116	15
	9	3	233	225	208	176	24
	10	2	279	275	258	223	46
	11	1	307	306	290	253	60
	Noon		317	317	301	263	65
	Daily total Btu/day-ft²		2585	2473	2281	1927	397
Jun 21	6	6	40	25	15	13	8
	7	5	111	91	75	49	13
	8	4	178	161	142	110	16
	9	3	235	222	203	167	17
	10	2	279	270	251	213	32
	11	1	307	300	282	242	46
	Noon		316	311	292	251	50
	Daily total Btu/day-ft²		2622	2456	2233	1846	319

Date	AM	PM	0°	4/12	Lat.	Lat. + 15°	90°
			Hourly clear day radiation, Btu/hr-ft²				
Jul 21	6	6	32	21	13	12	7
	7	5	104	88	73	51	13
	8	4	172	159	143	114	16
	9	3	230	222	205	173	24
	10	2	275	270	254	219	46
	11	1	303	301	285	249	60
	Noon		312	312	296	259	64
	Daily total Btu/day-ft²		2549	2438	2249	1901	400
Aug 21	6	6	11	8	5	5	3
	7	5	83	76	68	55	11
	8	4	155	153	144	125	29
	9	3	217	220	212	189	59
	10	2	265	273	265	239	83
	11	1	294	306	298	271	99
	Noon		305	317	310	282	104
	Daily total Btu/day-ft²		2362	2395	2300	2053	676
Sep 21	7	5	52	56	56	54	28
	8	4	127	140	141	135	71
	9	3	192	213	216	208	109
	10	2	242	270	274	263	138
	11	1	274	305	310	298	156
	Noon		285	318	323	310	162
	Daily total Btu/day-ft²		2062	2291	2323	2231	1169
Oct 21	7	5	21	30	34	37	34
	8	4	94	120	130	136	103
	9	3	160	197	210	215	150
	10	2	211	255	270	274	183
	11	1	243	292	308	311	203
	Noon		254	305	321	324	210
	Daily total Btu/day-ft²		1716	2096	2230	2276	1561
Nov 21	7	5	0	2	2	3	4
	8	4	61	93	108	122	115
	9	3	125	172	193	209	176
	10	2	175	231	255	271	214
	11	1	206	268	293	309	237
	Noon		217	280	306	322	244
	Daily total Btu/day-ft²		1356	1815	2015	2157	1739
Dec 21	7	5	0	0	0	0	0
	8	4	48	80	97	113	114
	9	3	110	161	185	205	184
	10	2	159	220	247	268	225
	11	1	190	257	286	306	248
	Noon		201	269	298	319	256
	Daily total Btu/day-ft²		1220	1708	1931	2107	1801

Table 26. Hourly, clear day solar radiation, continued.

32° North Latitude

Date	AM	PM	0°	4/12	Lat.	Lat. +15°	90°	Date	AM	PM	0°	4/12	Lat.	Lat. +15°	90°

Collector tilt angle — Hourly clear day radiation, Btu/hr-ft²

Date	AM	PM	0°	4/12	Lat.	Lat. +15°	90°
Jan 21	7	5	0	0	0	0	0
	8	4	56	87	106	119	115
	9	3	118	167	193	209	180
	10	2	167	226	256	272	221
	11	1	198	263	294	310	244
	Noon		209	275	307	323	252
Daily total Btu/day-ft²			1291	1767	2009	2150	1775
Feb 21	7	5	20	29	34	38	36
	8	4	92	120	133	140	109
	9	3	158	198	215	220	159
	10	2	209	257	277	281	194
	11	1	241	294	315	318	216
	Noon		252	307	328	331	223
Daily total Btu/day-ft²			1695	2108	2281	2330	1654
Mar 21	7	5	53	58	59	56	31
	8	4	128	144	146	140	78
	9	3	194	219	223	215	118
	10	2	244	276	283	272	150
	11	1	276	313	320	308	170
	Noon		287	325	333	320	176
Daily total Btu/day-ft²			2082	2350	2401	2307	1273
Apr 21	6	6	14	9	6	5	3
	7	5	86	80	70	56	9
	8	4	158	157	147	127	33
	9	3	220	226	217	193	66
	10	2	267	279	271	245	93
	11	1	297	313	305	278	110
	Noon		307	325	317	289	116
Daily total Btu/day-ft²			2397	2460	2353	2101	751
May 21	6	6	36	23	13	12	7
	7	5	107	91	74	51	12
	8	4	175	163	144	115	15
	9	3	233	227	208	175	31
	10	2	277	276	258	222	54
	11	1	305	307	289	252	70
	Noon		315	318	300	262	75
Daily total Btu/day-ft²			2587	2498	2279	1923	458
Jun 21	5	7	0	0	0	0	0
	6	6	44	28	16	14	8
	7	5	114	95	75	49	13
	8	4	179	163	142	110	16
	9	3	235	225	203	167	19
	10	2	278	272	251	212	41
	11	1	305	302	281	241	55
	Noon		314	312	291	251	60
Daily total Btu/day-ft²			2631	2488	2231	1844	369
Jul 21	6	6	35	23	14	12	7
	7	5	106	90	74	52	13
	8	4	173	161	143	114	16
	9	3	230	223	205	172	31
	10	2	273	272	254	219	54
	11	1	301	302	285	248	69
	Noon		310	313	295	258	74
Daily total Btu/day-ft²			2552	2463	2247	1897	460
Aug 21	6	6	13	9	6	5	3
	7	5	84	78	68	55	11
	8	4	154	153	143	124	33
	9	3	215	220	211	188	65
	10	2	261	272	264	238	91
	11	1	291	305	297	270	107
	Noon		301	317	309	281	113
Daily total Btu/day-ft²			2343	2399	2294	2046	740
Sep 21	7	5	50	55	55	53	29
	8	4	123	138	140	134	74
	9	3	187	210	215	206	114
	10	2	236	267	272	262	144
	11	1	267	302	309	297	163
	Noon		278	314	321	309	170
Daily total Btu/day-ft²			2012	2262	2306	2215	1224
Oct 21	7	5	18	26	30	33	31
	8	4	89	115	127	133	103
	9	3	153	191	207	212	153
	10	2	203	249	268	272	188
	11	1	234	286	306	309	209
	Noon		245	298	319	321	217
Daily total Btu/day-ft²			1644	2036	2200	2244	1590
Nov 21	7	5	0	0	0	0	0
	8	4	54	85	103	116	111
	9	3	117	164	189	205	176
	10	2	165	223	252	268	217
	11	1	196	259	290	306	240
	Noon		206	272	303	318	248
Daily total Btu/day-ft²			1273	1739	1974	2109	1739
Dec 21	8	4	41	71	89	104	106
	9	3	101	152	180	199	182
	10	2	149	211	243	264	226
	11	1	179	247	282	302	251
	Noon		189	259	295	315	259
Daily total Btu/day-ft²			1133	1624	1886	2057	1792

Table 26. Hourly, clear day solar radiation, continued.

34° North Latitude

Date	AM	PM	0°	4/12	Lat.	Lat. + 15°	90°
			Hourly clear day radiation, Btu/hr-ft²				
Jan 21	7	5	0	0	0	0	0
	8	4	49	79	99	112	109
	9	3	110	159	188	204	179
	10	2	157	218	252	268	222
	11	1	188	254	291	307	248
	Noon		198	266	304	320	256
	Daily total Btu/day-ft²		1208	1689	1968	2105	1776
Feb 21	7	5	16	25	31	34	32
	8	4	87	115	130	136	109
	9	3	151	192	212	218	162
	10	2	200	251	274	278	199
	11	1	232	288	313	316	222
	Noon		242	300	326	329	230
	Daily total Btu/day-ft²		1618	2046	2251	2298	1682
Mar 21	7	5	51	57	58	55	32
	8	4	125	141	145	139	81
	9	3	189	216	222	213	124
	10	2	238	273	281	270	157
	11	1	269	309	318	306	178
	Noon		280	321	331	319	185
	Daily total Btu/day-ft²		2028	2317	2384	2292	1330
Apr 21	6	6	15	11	6	5	3
	7	5	87	81	70	55	9
	8	4	157	158	147	127	38
	9	3	217	226	216	192	73
	10	2	264	279	270	244	101
	11	1	293	312	304	276	119
	Noon		303	323	316	288	125
	Daily total Btu/day-ft²		2374	2459	2346	2093	817
May 21	6	6	39	26	14	12	7
	7	5	109	94	75	51	12
	8	4	175	165	144	115	15
	9	3	232	228	207	174	38
	10	2	275	277	257	221	63
	11	1	303	308	289	251	79
	Noon		312	318	299	262	84
	Daily total Btu/day-ft²		2585	2519	2276	1919	517
Jun 21	5	7	0	0	0	0	0
	6	6	48	32	16	15	9
	7	5	116	98	75	49	14
	8	4	180	166	142	110	16
	9	3	235	227	203	167	26
	10	2	277	274	250	211	49
	11	1	303	303	280	240	64
	Noon		312	313	291	250	69
	Daily total Btu/day-ft²		2638	2517	2230	1840	429

Date	AM	PM	0°	4/12	Lat.	Lat. + 15°	90°
			Hourly clear day radiation, Btu/hr-ft²				
Jul 21	6	6	39	26	14	13	8
	7	5	108	93	74	52	13
	8	4	173	163	142	114	16
	9	3	229	225	204	172	38
	10	2	272	273	253	218	62
	11	1	299	303	284	247	78
	Noon		308	313	295	257	83
	Daily total Btu/day-ft²		2552	2484	2245	1892	518
Aug 21	6	6	14	10	6	6	3
	7	5	84	79	68	55	11
	8	4	153	154	143	123	38
	9	3	213	220	210	187	71
	10	2	258	272	263	237	98
	11	1	286	304	296	269	116
	Noon		296	315	307	280	122
	Daily total Btu/day-ft²		2320	2397	2285	2038	803
Sep 21	7	5	48	53	54	51	30
	8	4	120	135	138	132	77
	9	3	182	207	213	204	119
	10	2	230	263	270	260	151
	11	1	261	298	307	295	171
	Noon		271	310	319	307	178
	Daily total Btu/day-ft²		1959	2229	2288	2197	1276
Oct 21	7	5	15	22	26	29	27
	8	4	83	110	124	129	103
	9	3	146	185	204	209	155
	10	2	195	243	265	269	192
	11	1	226	279	303	306	215
	Noon		236	292	316	319	223
	Daily total Btu/day-ft²		1570	1975	2167	2209	1614
Nov 21	7	5	0	0	0	0	0
	8	4	47	77	96	108	105
	9	3	108	156	184	200	175
	10	2	155	214	248	263	218
	11	1	185	250	286	302	243
	Noon		196	263	299	315	252
	Daily total Btu/day-ft²		1191	1661	1931	2064	1739
Dec 21	8	4	33	61	80	93	97
	9	3	92	142	174	193	179
	10	2	139	201	238	259	226
	11	1	168	237	278	298	252
	Noon		178	250	291	311	261
	Daily total Btu/day-ft²		1046	1537	1835	1999	1773

Table 26. Hourly, clear day solar radiation, continued.

36° North Latitude

Date	AM	PM	0°	$\frac{4}{12}$	Lat.	Lat. + 15°	90°	Date	AM	PM	0°	$\frac{4}{12}$	Lat.	Lat. + 15°	90°
Jan 21	8	4	42	71	92	104	102	Jul 21	5	7	0	0	0	0	0
	9	3	101	150	183	199	177		6	6	42	29	15	13	8
	10	2	147	208	247	263	223		7	5	110	96	74	52	13
	11	1	177	245	287	302	250		8	4	173	165	142	113	16
	Noon		187	257	300	315	259		9	3	228	226	204	171	45
Daily total									10	2	270	273	252	217	70
Btu/day-ft²			1123	1607	1922	2054	1769		11	1	296	303	283	246	87
Feb 21	7	5	13	21	26	29	28		Noon		305	313	294	256	92
	8	4	81	110	127	133	109	Daily total							
	9	3	144	186	209	215	164	Btu/day-ft²			2547	2501	2240	1886	575
	10	2	192	244	271	275	203	Aug 21	6	6	16	12	7	6	4
	11	1	222	281	310	313	228		7	5	85	80	68	54	11
	Noon		233	293	323	326	236		8	4	152	154	142	123	42
Daily total									9	3	210	219	209	186	77
Btu/day-ft²			1542	1980	2216	2261	1704		10	2	254	270	261	236	106
Mar 21	7	5	49	56	57	54	33		11	1	282	302	295	267	124
	8	4	121	139	144	138	84		Noon		291	313	306	278	131
	9	3	183	212	220	211	129	Daily total							
	10	2	232	269	279	269	163	Btu/day-ft²			2294	2393	2276	2027	865
	11	1	262	304	317	305	185	Sep 21	7	5	46	52	53	50	30
	Noon		272	317	329	317	193		8	4	116	133	137	131	79
Daily total									9	3	177	204	211	202	123
Btu/day-ft²			1971	2281	2367	2274	1384		10	2	224	259	269	258	157
Apr 21	6	6	17	12	7	6	3		11	1	254	294	305	293	178
	7	5	87	82	70	55	9		Noon		264	305	317	305	185
	8	4	155	158	146	126	42	Daily total							
	9	3	214	225	215	191	79	Btu/day-ft²			1904	2192	2269	2177	1325
	10	2	260	277	269	243	109	Oct 21	7	5	12	18	22	25	23
	11	1	288	310	303	275	128		8	4	78	104	120	126	102
	Noon		298	322	315	286	134		9	3	139	179	201	206	157
Daily total									10	2	187	236	262	266	196
Btu/day-ft²			2347	2455	2338	2084	882		11	1	216	272	300	303	220
May 21	5	7	0	0	0	0	0		Noon		227	284	313	316	228
	6	6	43	29	14	13	8	Daily total							
	7	5	111	97	75	51	12	Btu/day-ft²			1494	1908	2131	2171	1632
	8	4	175	167	144	115	15	Nov 21	8	4	41	68	88	100	98
	9	3	231	229	207	174	45		9	3	99	147	179	194	173
	10	2	273	277	256	221	71		10	2	145	205	243	258	219
	11	1	300	308	288	250	88		11	1	175	241	282	297	246
	Noon		309	318	299	261	94		Noon		185	253	295	310	254
Daily total								Daily total							
Btu/day-ft²			2581	2537	2272	1914	577	Btu/day-ft²			1108	1578	1884	2012	1729
Jun 21	5	7	0	0	0	0	0	Dec 21	8	4	26	51	70	81	85
	6	6	52	35	17	15	9		9	3	83	132	167	186	175
	7	5	119	101	76	49	14		10	2	128	191	233	253	225
	8	4	181	168	142	110	16		11	1	157	227	273	292	253
	9	3	234	228	202	166	33		Noon		167	239	286	306	263
	10	2	275	275	249	211	57	Daily total							
	11	1	301	304	280	239	73	Btu/day-ft²			959	1444	1776	1934	1744
	Noon		310	314	290	249	79								
Daily total															
Btu/day-ft²			2642	2543	2227	1836	489								

Collector tilt angle — Hourly clear day radiation, Btu/hr-ft²

Table 26. Hourly, clear day solar radiation, continued.

38° North Latitude

Date	AM	PM	0°	4/12	Lat.	Lat. + 15°	90°
							Hourly clear day radiation, Btu/hr-ft²
Jan 21	8	4	35	61	83	94	94
	9	3	92	141	177	192	175
	10	2	137	199	243	258	224
	11	1	166	235	282	298	252
	Noon		175	247	296	311	261
Daily total Btu/day-ft²			1037	1521	1871	1998	1752
Feb 21	7	5	10	17	22	24	24
	8	4	75	104	123	129	108
	9	3	136	179	206	211	166
	10	2	183	237	269	272	207
	11	1	213	273	307	310	233
	Noon		223	285	320	323	241
Daily total Btu/day-ft²			1462	1910	2179	2222	1720
Mar 21	7	5	47	54	56	53	33
	8	4	117	136	142	136	86
	9	3	178	208	218	210	133
	10	2	225	264	277	267	170
	11	1	254	299	315	303	192
	Noon		264	311	328	315	200
Daily total Btu/day-ft²			1911	2240	2349	2255	1434
Apr 21	6	6	19	14	7	6	4
	7	5	87	83	70	55	10
	8	4	154	158	145	125	47
	9	3	211	224	214	190	85
	10	2	255	276	268	241	116
	11	1	283	308	302	274	136
	Noon		293	319	313	285	143
Daily total Btu/day-ft²			2317	2447	2328	2074	945
May 21	5	7	0	0	0	0	0
	6	6	46	32	15	13	8
	7	5	113	99	75	51	12
	8	4	175	168	144	114	18
	9	3	229	230	206	173	52
	10	2	271	277	255	220	79
	11	1	297	308	287	249	97
	Noon		305	318	298	260	103
Daily total Btu/day-ft²			2571	2552	2267	1907	642
Jun 21	5	7	1	1	1	1	0
	6	6	56	39	17	15	9
	7	5	121	104	76	49	14
	8	4	182	171	142	109	16
	9	3	234	230	202	165	40
	10	2	273	275	249	210	66
	11	1	299	305	279	238	82
	Noon		307	314	289	248	88
Daily total Btu/day-ft²			2644	2567	2225	1831	548

Date	AM	PM	0°	4/12	Lat.	Lat. + 15°	90°
							Hourly clear day radiation, Btu/hr-ft²
Jul 21	5	7	0	0	0	0	0
	6	6	46	32	15	14	8
	7	5	111	98	75	51	14
	8	4	173	166	142	113	19
	9	3	226	226	203	170	51
	10	2	267	273	251	216	78
	11	1	293	303	282	245	96
	Noon		301	313	293	255	102
Daily total Btu/day-ft²			2539	2517	2235	1880	639
Aug 21	6	6	18	13	7	7	4
	7	5	85	81	68	54	11
	8	4	150	154	141	122	46
	9	3	207	218	208	185	83
	10	2	250	269	260	234	113
	11	1	277	300	293	266	133
	Noon		286	311	305	277	139
Daily total Btu/day-ft²			2265	2386	2267	2017	924
Sep 21	7	5	45	50	51	49	31
	8	4	112	130	135	129	82
	9	3	172	200	209	200	127
	10	2	218	254	266	256	163
	11	1	246	289	303	291	185
	Noon		256	300	315	303	192
Daily total Btu/day-ft²			1846	2152	2249	2157	1371
Oct 21	7	5	9	14	18	20	19
	8	4	72	98	116	121	101
	9	3	132	173	198	203	159
	10	2	178	229	259	263	200
	11	1	207	265	297	300	225
	Noon		217	277	310	313	233
Daily total Btu/day-ft²			1416	1840	2092	2130	1645
Nov 21	8	4	34	59	80	90	89
	9	3	90	138	173	187	170
	10	2	135	195	238	253	219
	11	1	164	231	278	292	247
	Noon		173	243	291	306	256
Daily total Btu/day-ft²			1022	1493	1832	1955	1711
Dec 21	8	4	19	40	58	67	71
	9	3	74	122	160	177	169
	10	2	117	180	227	246	223
	11	1	145	216	267	287	253
	Noon		155	228	281	300	263
Daily total Btu/day-ft²			870	1347	1709	1858	1701

Table 26. Hourly, clear day solar radiation, continued.

40° North Latitude

Date	AM	PM	0°	⁴/₁₂	Lat.	Lat. + 15°	90°
Jan 21	8	4	28	52	74	83	84
	9	3	83	131	171	185	171
	10	2	127	189	237	252	223
	11	1	154	224	277	292	252
	Noon		164	236	291	306	262
Daily total Btu/day-ft²			951	1431	1812	1934	1725
Feb 21	7	5	7	13	17	19	19
	8	4	69	98	119	125	107
	9	3	128	173	203	208	167
	10	2	174	229	265	269	211
	11	1	203	265	304	307	237
	Noon		213	277	317	320	246
Daily total Btu/day-ft²			1382	1837	2139	2180	1731
Mar 21	7	5	45	52	54	52	34
	8	4	113	133	140	134	89
	9	3	172	204	216	208	138
	10	2	218	259	275	265	175
	11	1	246	294	313	301	199
	Noon		256	306	325	313	207
Daily total Btu/day-ft²			1849	2196	2327	2235	1481
Apr 21	6	6	20	15	7	6	4
	7	5	87	83	70	55	10
	8	4	152	157	144	124	51
	9	3	208	223	213	189	91
	10	2	251	273	266	240	124
	11	1	278	306	300	273	145
	Noon		287	317	312	284	152
Daily total Btu/day-ft²			2284	2437	2318	2064	1008
May 21	5	7	0	0	0	0	0
	6	6	49	35	15	13	8
	7	5	114	101	75	51	13
	8	4	175	170	143	114	23
	9	3	227	230	205	172	59
	10	2	268	277	255	219	87
	11	1	293	307	286	248	106
	Noon		301	317	296	259	112
Daily total Btu/day-ft²			2560	2563	2260	1900	710
Jun 21	5	7	4	2	2	2	1
	6	6	60	42	18	16	10
	7	5	123	107	76	49	14
	8	4	182	172	141	109	16
	9	3	232	231	201	165	46
	10	2	271	276	248	209	74
	11	1	296	305	278	237	91
	Noon		304	314	288	247	97
Daily total Btu/day-ft²			2644	2589	2222	1827	608

Date	AM	PM	0°	⁴/₁₂	Lat.	Lat. + 15°	90°
Jul 21	5	7	0	0	0	0	0
	6	6	49	35	16	14	9
	7	5	113	100	75	51	14
	8	4	173	167	141	112	24
	9	3	225	227	202	170	58
	10	2	264	273	250	215	86
	11	1	289	302	281	244	104
	Noon		297	312	291	254	110
Daily total Btu/day-ft²			2528	2528	2227	1872	704
Aug 21	6	6	20	15	8	7	4
	7	5	85	81	68	54	11
	8	4	149	153	140	121	50
	9	3	203	217	207	183	89
	10	2	245	267	259	233	120
	11	1	272	298	292	264	140
	Noon		281	309	303	275	147
Daily total Btu/day-ft²			2234	2376	2255	2005	983
Sep 21	7	5	43	49	50	47	31
	8	4	109	127	133	127	84
	9	3	166	196	207	198	131
	10	2	211	250	264	253	168
	11	1	239	283	300	288	191
	Noon		248	295	313	300	199
Daily total Btu/day-ft²			1786	2108	2226	2133	1413
Oct 21	7	5	6	10	14	15	15
	8	4	66	93	112	117	100
	9	3	124	166	194	199	160
	10	2	169	222	256	259	203
	11	1	197	257	294	296	229
	Noon		207	269	307	309	237
Daily total Btu/day-ft²			1337	1767	2050	2086	1652
Nov 21	8	4	27	50	70	79	79
	9	3	81	128	166	180	166
	10	2	125	185	232	247	218
	11	1	152	221	272	287	248
	Noon		162	233	286	300	257
Daily total Btu/day-ft²			936	1403	1772	1889	1682
Dec 21	8	4	13	29	44	51	55
	9	3	64	111	151	167	162
	10	2	107	169	220	239	220
	11	1	134	204	261	280	252
	Noon		143	216	275	294	262
Daily total Btu/day-ft²			781	1245	1632	1772	1643

The column headers span "Collector tilt angle" with subtitle "Hourly clear day radiation, Btu/hr-ft²".

Table 26. Hourly, clear day solar radiation, continued.

42° North Latitude

Date	AM	PM	Collector tilt angle 0°	4/12	Lat.	Lat. + 15°	90°		Date	AM	PM	Collector tilt angle 0°	4/12	Lat.	Lat. + 15°	90°
			Hourly clear day radiation, Btu/hr-ft^2									Hourly clear day radiation, Btu/hr-ft^2				
Jan 21	8	4	21	42	62	70	71		Jul 21	5	7	1	0	0	0	0
	9	3	74	121	163	177	166			6	6	52	38	16	14	9
	10	2	116	178	231	246	221			7	5	114	103	75	51	14
	11	1	143	213	272	286	252			8	4	173	169	141	112	29
	Noon		152	225	285	300	262			9	3	223	227	201	169	64
	Daily total									10	2	261	272	249	214	94
	Btu/day-ft^2		863	1336	1744	1861	1685			11	1	285	301	280	243	113
Feb 21	7	5	5	9	12	14	14			Noon		293	311	290	253	119
	8	4	63	92	115	120	105			Daily total						
	9	3	120	165	199	204	168			Btu/day-ft^2		2515	2537	2221	1865	769
	10	2	165	221	262	265	213		Aug 21	6	6	21	16	8	7	5
	11	1	193	257	301	303	241			7	5	85	82	68	53	14
	Noon		202	269	314	316	250			8	4	147	153	139	120	54
	Daily total									9	3	200	215	205	182	94
	Btu/day-ft^2		1299	1761	2095	2133	1734			10	2	240	264	257	231	127
Mar 21	7	5	43	50	53	50	34			11	1	266	295	290	263	148
	8	4	109	130	138	132	91			Noon		275	305	301	274	155
	9	3	166	200	214	206	141			Daily total						
	10	2	210	254	273	262	181			Btu/day-ft^2		2198	2360	2242	1993	1043
	11	1	238	288	310	298	206		Sep 21	7	5	41	47	48	46	31
	Noon		248	300	323	311	214			8	4	104	123	131	125	86
	Daily total									9	3	160	192	205	196	135
	Btu/day-ft^2		1785	2149	2305	2213	1524			10	2	203	244	262	251	173
Apr 21	6	6	22	17	8	7	4			11	1	230	278	298	286	197
	7	5	87	84	69	54	13			Noon		240	289	310	298	205
	8	4	150	157	144	124	55			Daily total						
	9	3	204	221	211	188	97			Btu/day-ft^2		1723	2061	2201	2110	1452
	10	2	246	271	265	239	131		Oct 21	7	5	4	7	9	10	10
	11	1	272	303	299	271	153			8	4	60	86	107	112	97
	Noon		281	313	310	282	160			9	3	116	159	190	194	160
	Daily total									10	2	160	214	252	255	205
	Btu/day-ft^2		2248	2422	2306	2052	1072			11	1	187	248	290	293	232
May 21	5	7	1	0	0	0	0			Noon		197	260	303	305	241
	6	6	52	38	15	13	9			Daily total						
	7	5	115	104	75	51	13			Btu/day-ft^2		1256	1692	2004	2038	1653
	8	4	175	171	143	113	28		Nov 21	8	4	20	40	59	66	67
	9	3	225	230	205	171	65			9	3	72	118	159	172	161
	10	2	264	277	253	218	95			10	2	114	175	226	240	216
	11	1	289	306	285	247	115			11	1	141	210	267	281	247
	Noon		297	316	295	257	121			Noon		150	222	280	294	257
	Daily total									Daily total						
	Btu/day-ft^2		2546	2573	2254	1892	777			Btu/day-ft^2		849	1309	1704	1815	1641
Jun 21	5	7	7	4	4	3	2		Dec 21	8	4	7	18	29	33	36
	6	6	63	46	18	16	10			9	3	55	99	141	156	153
	7	5	124	109	76	49	14			10	2	96	157	212	230	215
	8	4	182	174	141	109	19			11	1	122	192	254	273	249
	9	3	231	231	200	164	53			Noon		131	204	268	287	260
	10	2	269	276	247	208	82			Daily total						
	11	1	292	304	277	236	100			Btu/day-ft^2		693	1139	1544	1675	1570
	Noon		300	314	287	246	106									
	Daily total															
	Btu/day-ft^2		2643	2609	2219	1822	672									

Table 26. Hourly, clear day solar radiation, continued.

44° North Latitude

Date	AM	PM	0°	$^4/_{12}$	Lat.	Lat. + 15°	90°
				Collector tilt angle			
				Hourly clear day radiation, Btu/hr-ft²			
Jan 21	8	4	15	31	49	56	57
	9	3	64	110	154	167	159
	10	2	105	167	224	238	217
	11	1	131	202	265	279	250
	Noon		140	214	279	293	261
	Daily total Btu/day-ft²		775	1237	1668	1777	1632
Feb 21	7	5	3	5	8	9	9
	8	4	57	85	110	115	102
	9	3	112	158	195	199	168
	10	2	155	213	258	261	215
	11	1	182	248	297	300	244
	Noon		191	259	310	313	253
	Daily total Btu/day-ft²		1215	1681	2048	2084	1732
Mar 21	7	5	41	48	51	49	34
	8	4	104	126	136	130	93
	9	3	160	195	212	203	145
	10	2	203	248	271	260	186
	11	1	230	282	308	296	212
	Noon		239	293	321	308	220
	Daily total Btu/day-ft²		1718	2097	2281	2188	1564
Apr 21	6	6	24	18	8	7	4
	7	5	87	85	69	54	15
	8	4	148	156	143	123	59
	9	3	200	219	210	187	103
	10	2	240	268	263	237	138
	11	1	266	299	297	269	160
	Noon		274	310	309	280	168
	Daily total Btu/day-ft²		2208	2405	2294	2039	1134
May 21	5	7	3	2	1	1	1
	6	6	55	41	16	14	9
	7	5	116	106	75	51	13
	8	4	174	172	142	113	33
	9	3	223	230	204	171	72
	10	2	261	276	252	217	103
	11	1	284	304	283	246	123
	Noon		292	314	294	256	130
	Daily total Btu/day-ft²		2531	2579	2247	1885	843
Jun 21	5	7	11	6	5	4	3
	6	6	67	49	18	16	10
	7	5	126	112	76	49	14
	8	4	182	175	141	108	24
	9	3	229	232	199	163	60
	10	2	265	276	246	207	90
	11	1	288	303	276	235	109
	Noon		296	313	286	245	115
	Daily total Btu/day-ft²		2639	2625	2214	1817	740

Date	AM	PM	0°	$^4/_{12}$	Lat.	Lat. + 15°	90°
				Collector tilt angle			
				Hourly clear day radiation, Btu/hr-ft²			
Jul 21	5	7	3	2	1	1	1
	6	6	55	41	17	15	9
	7	5	115	105	75	51	14
	8	4	172	169	140	111	33
	9	3	220	227	200	168	71
	10	2	257	271	248	213	101
	11	1	281	300	278	241	121
	Noon		289	309	289	251	128
	Daily total Btu/day-ft²		2501	2544	2215	1857	834
Aug 21	6	6	23	18	9	8	5
	7	5	85	82	67	53	16
	8	4	144	152	138	119	58
	9	3	196	213	204	181	100
	10	2	235	261	255	230	133
	11	1	260	291	288	261	155
	Noon		268	302	299	272	163
	Daily total Btu/day-ft²		2160	2343	2228	1979	1102
Sep 21	7	5	39	45	47	44	31
	8	4	100	120	128	123	87
	9	3	154	187	202	193	138
	10	2	196	239	259	248	177
	11	1	222	271	295	283	202
	Noon		231	283	307	295	211
	Daily total Btu/day-ft²		1657	2011	2175	2083	1487
Oct 21	7	5	2	4	5	6	6
	8	4	54	80	102	106	94
	9	3	108	151	185	189	160
	10	2	150	205	247	251	206
	11	1	177	239	286	288	234
	Noon		186	251	299	301	244
	Daily total Btu/day-ft²		1175	1614	1956	1987	1648
Nov 21	8	4	14	30	46	52	53
	9	3	63	107	150	162	154
	10	2	104	163	219	232	212
	11	1	129	198	260	274	245
	Noon		138	210	274	287	256
	Daily total Btu/day-ft²		762	1209	1626	1730	1587
Dec 21	8	4	2	7	13	15	17
	9	3	45	86	129	142	141
	10	2	85	145	203	220	209
	11	1	110	180	246	264	245
	Noon		118	192	260	278	257
	Daily total Btu/day-ft²		605	1032	1446	1565	1483

Table 26. Hourly, clear day solar radiation, continued.

46° North Latitude

Date	AM	PM	0°	4/12	Lat.	Lat. + 15°	90°
					Collector tilt angle		
			Hourly clear day radiation, Btu/hr-ft²				
Jan 21	8	4	9	21	35	39	41
	9	3	55	98	144	156	150
	10	2	94	155	216	229	213
	11	1	119	190	258	272	248
	Noon		128	201	272	286	259
	Daily total Btu/day-ft²		687	1132	1580	1682	1564
Feb 21	7	5	1	2	4	4	4
	8	4	51	78	104	109	98
	9	3	104	150	190	194	167
	10	2	145	204	253	257	216
	11	1	172	238	293	295	246
	Noon		180	250	306	308	256
	Daily total Btu/day-ft²		1132	1601	1997	2031	1724
Mar 21	7	5	39	46	50	47	34
	8	4	100	122	134	128	94
	9	3	153	190	209	201	148
	10	2	195	242	268	257	190
	11	1	221	275	305	293	217
	Noon		229	287	318	305	226
	Daily total Btu/day-ft²		1648	2043	2254	2162	1599
Apr 21	6	6	25	20	8	7	5
	7	5	86	85	69	54	17
	8	4	145	155	142	122	63
	9	3	196	217	209	185	108
	10	2	235	265	262	236	145
	11	1	259	295	295	268	168
	Noon		267	306	307	279	176
	Daily total Btu/day-ft²		2165	2383	2280	2026	1194
May 21	5	7	6	3	3	2	1
	6	6	58	44	16	14	9
	7	5	117	107	75	51	13
	8	4	173	172	142	112	38
	9	3	220	229	203	170	78
	10	2	257	274	251	215	110
	11	1	279	302	282	245	131
	Noon		287	312	292	255	139
	Daily total Btu/day-ft²		2514	2583	2240	1878	908
Jun 21	5	7	16	8	7	6	4
	6	6	70	52	19	16	11
	7	5	127	114	76	49	14
	8	4	181	177	140	107	29
	9	3	227	232	198	162	67
	10	2	262	275	245	206	97
	11	1	284	302	274	234	117
	Noon		292	311	284	243	124
	Daily total Btu/day-ft²		2633	2636	2208	1809	807

Date	AM	PM	0°	4/12	Lat.	Lat. + 15°	90°
					Collector tilt angle		
			Hourly clear day radiation, Btu/hr-ft²				
Jul 21	5	7	6	3	3	2	1
	6	6	58	44	17	15	10
	7	5	116	106	74	51	14
	8	4	171	170	140	111	38
	9	3	218	226	199	167	77
	10	2	253	270	247	211	108
	11	1	276	298	277	240	129
	Noon		284	307	287	250	136
	Daily total Btu/day-ft²		2485	2548	2206	1849	897
Aug 21	6	6	24	19	9	8	5
	7	5	84	83	67	53	18
	8	4	142	151	137	118	62
	9	3	192	211	202	179	105
	10	2	230	258	254	228	140
	11	1	254	288	286	259	162
	Noon		262	298	297	270	170
	Daily total Btu/day-ft²		2119	2321	2213	1964	1158
Sep 21	7	5	36	43	45	43	31
	8	4	96	116	126	120	88
	9	3	148	182	199	191	141
	10	2	188	233	256	246	181
	11	1	213	265	292	280	207
	Noon		222	276	304	292	216
	Daily total Btu/day-ft²		1589	1956	2146	2055	1517
Oct 21	7	5	1	1	2	3	3
	8	4	48	73	96	100	90
	9	3	100	143	180	184	158
	10	2	141	197	243	246	207
	11	1	166	230	281	284	236
	Noon		175	241	294	296	246
	Daily total Btu/day-ft²		1093	1533	1903	1933	1638
Nov 21	8	4	8	19	32	36	37
	9	3	54	95	139	151	145
	10	2	93	152	210	224	207
	11	1	118	186	252	266	242
	Noon		126	198	266	280	253
	Daily total Btu/day-ft²		675	1106	1537	1635	1518
Dec 21	8	4	0	1	2	2	2
	9	3	36	73	114	126	126
	10	2	73	131	192	209	200
	11	1	98	166	237	254	239
	Noon		106	178	251	269	251
	Daily total Btu/day-ft²		523	924	1345	1455	1390

Table 26. Hourly, clear day solar radiation, continued.

48° North Latitude

Date	AM	PM	Collector tilt angle 0°	4/12	Lat.	Lat. + 15°	90°
			Hourly clear day radiation, Btu/hr-ft²				
Jan 21	8	4	4	11	19	22	23
	9	3	46	86	132	143	139
	10	2	83	143	206	219	206
	11	1	107	177	250	263	243
	Noon		116	188	264	277	255
	Daily total Btu/day-ft²		601	1024	1482	1574	1480
Feb 21	7	5	0	0	1	1	1
	8	4	45	71	98	102	94
	9	3	96	141	184	189	166
	10	2	136	195	248	252	216
	11	1	161	228	288	290	247
	Noon		169	240	301	303	257
	Daily total Btu/day-ft²		1048	1517	1944	1975	1709
Mar 21	7	5	37	44	48	45	34
	8	4	95	118	131	125	95
	9	3	147	184	207	198	151
	10	2	186	236	265	255	194
	11	1	211	268	302	290	222
	Noon		220	279	315	303	231
	Daily total Btu/day-ft²		1576	1984	2226	2134	1629
Apr 21	6	6	27	21	8	7	5
	7	5	86	85	68	53	20
	8	4	142	153	140	121	67
	9	3	191	214	207	184	114
	10	2	229	261	260	234	151
	11	1	252	291	293	266	175
	Noon		260	301	305	277	183
	Daily total Btu/day-ft²		2120	2357	2265	2011	1251
May 21	5	7	9	5	4	3	2
	6	6	61	47	16	14	9
	7	5	118	109	75	50	13
	8	4	171	172	141	111	43
	9	3	217	229	202	168	84
	10	2	252	272	250	214	118
	11	1	274	300	280	243	139
	Noon		282	309	291	253	147
	Daily total Btu/day-ft²		2495	2583	2232	1868	972
Jun 21	5	7	20	9	8	7	5
	6	6	73	56	19	16	11
	7	5	129	116	76	48	14
	8	4	180	178	139	107	34
	9	3	224	232	198	161	73
	10	2	258	274	244	205	105
	11	1	280	300	273	232	125
	Noon		287	309	283	242	132
	Daily total Btu/day-ft²		2623	2644	2202	1802	872

Date	AM	PM	Collector tilt angle 0°	4/12	Lat.	Lat. + 15°	90°
			Hourly clear day radiation, Btu/hr-ft²				
Jul 21	5	7	9	5	4	4	2
	6	6	61	47	17	15	10
	7	5	117	108	74	50	14
	8	4	170	170	139	110	43
	9	3	215	225	198	165	83
	10	2	249	268	245	210	115
	11	1	271	295	275	238	136
	Noon		278	305	286	248	144
	Daily total Btu/day-ft²		2466	2549	2199	1839	958
Aug 21	6	6	26	21	9	8	5
	7	5	84	83	66	52	20
	8	4	139	149	136	117	65
	9	3	187	208	201	177	110
	10	2	224	254	252	226	146
	11	1	247	283	284	257	169
	Noon		255	293	295	268	177
	Daily total Btu/day-ft²		2075	2296	2196	1947	1211
Sep 21	7	5	34	41	43	41	31
	8	4	91	112	123	118	89
	9	3	141	176	196	188	143
	10	2	180	226	253	242	185
	11	1	204	258	289	277	211
	Noon		213	268	301	289	221
	Daily total Btu/day-ft²		1519	1898	2115	2024	1543
Oct 21	7	5	0	0	0	0	0
	8	4	42	66	89	93	85
	9	3	92	135	174	178	156
	10	2	131	187	237	240	206
	11	1	156	220	276	278	237
	Noon		164	231	289	291	247
	Daily total Btu/day-ft²		1010	1451	1849	1876	1621
Nov 21	8	4	4	10	17	19	20
	9	3	45	83	127	137	133
	10	2	82	139	201	213	200
	11	1	106	173	244	257	237
	Noon		114	185	258	271	249
	Daily total Btu/day-ft²		590	1000	1439	1528	1434
Dec 21	8	4	0	0	0	0	0
	9	3	27	59	97	107	108
	10	2	62	117	180	195	189
	11	1	85	152	226	242	231
	Noon		93	164	241	257	244
	Daily total Btu/day-ft²		444	823	1248	1347	1302

Table 26. Hourly, clear day solar radiation, continued.

50° North Latitude

Date	AM	PM	0°	4/12	Lat.	Lat. + 15°	90°
					Hourly clear day radiation, Btu/hr-ft²		
Jan 21	8	4	1	3	6	7	7
	9	3	37	73	118	128	125
	10	2	72	129	195	207	197
	11	1	95	163	240	252	237
	Noon		103	175	254	267	249
	Daily total Btu/day-ft²		517	917	1375	1459	1385
Feb 21	7	5	0	0	0	0	0
	8	4	39	64	91	95	88
	9	3	87	133	178	182	163
	10	2	125	185	243	246	216
	11	1	149	218	282	285	247
	Noon		158	229	296	298	258
	Daily total Btu/day-ft²		964	1431	1888	1918	1690
Mar 21	7	5	34	42	46	43	34
	8	4	90	114	128	123	96
	9	3	140	179	204	195	153
	10	2	178	229	262	251	198
	11	1	202	260	299	287	226
	Noon		210	271	312	299	236
	Daily total Btu/day-ft²		1502	1922	2194	2103	1655
Apr 21	6	6	29	23	9	7	5
	7	5	85	85	68	53	22
	8	4	139	152	139	119	71
	9	3	186	211	205	182	118
	10	2	222	257	258	232	157
	11	1	245	286	291	264	182
	Noon		253	296	303	275	190
	Daily total Btu/day-ft²		2072	2329	2248	1995	1305
May 21	5	7	13	6	5	4	3
	6	6	64	49	16	14	10
	7	5	119	111	74	50	13
	8	4	170	173	140	111	48
	9	3	214	227	200	167	90
	10	2	247	270	248	213	125
	11	1	269	297	279	242	147
	Noon		276	306	289	251	155
	Daily total Btu/day-ft²		2473	2579	2222	1858	1033
Jun 21	4	8	0	0	0	0	0
	5	7	25	12	9	8	5
	6	6	77	59	19	16	11
	7	5	130	119	76	48	14
	8	4	179	178	139	106	39
	9	3	222	231	196	160	79
	10	2	254	272	242	203	112
	11	1	275	298	271	231	133
	Noon		282	307	281	241	140
	Daily total Btu/day-ft²		2610	2650	2193	1793	936

Date	AM	PM	0°	4/12	Lat.	Lat. + 15°	90°
					Hourly clear day radiation, Btu/hr-ft²		
Jul 21	5	7	13	7	6	5	3
	6	6	64	49	17	15	10
	7	5	118	110	74	50	14
	8	4	168	170	138	109	48
	9	3	211	224	197	164	89
	10	2	244	266	244	208	122
	11	1	265	293	274	237	144
	Noon		272	302	284	247	151
	Daily total Btu/day-ft²		2445	2545	2188	1829	1017
Aug 21	6	6	28	22	9	8	6
	7	5	83	83	66	51	22
	8	4	137	148	135	115	68
	9	3	182	205	199	176	114
	10	2	218	250	249	224	151
	11	1	240	279	282	255	175
	Noon		247	288	293	265	183
	Daily total Btu/day-ft²		2027	2267	2178	1929	1261
Sep 21	7	5	32	38	41	39	30
	8	4	86	108	120	115	90
	9	3	135	171	193	184	145
	10	2	172	219	250	239	188
	11	1	195	250	285	274	215
	Noon		203	260	298	285	225
	Daily total Btu/day-ft²		1447	1836	2081	1991	1564
Oct 21	7	5	0	0	0	0	0
	8	4	36	58	82	86	79
	9	3	84	126	168	172	153
	10	2	121	177	231	234	205
	11	1	145	210	270	273	237
	Noon		153	221	284	285	247
	Daily total Btu/day-ft²		928	1367	1791	1816	1598
Nov 21	8	4	1	2	5	6	6
	9	3	36	71	113	122	119
	10	2	71	126	189	201	191
	11	1	94	160	234	246	230
	Noon		102	171	248	261	243
	Daily total Btu/day-ft²		507	892	1333	1414	1340
Dec 21	9	3	18	43	76	84	85
	10	2	51	102	164	178	175
	11	1	73	137	213	228	220
	Noon		81	149	229	244	234
	Daily total Btu/day-ft²		368	717	1138	1227	1196

Table 27. Average total daily solar radiation.

South-facing collectors near selected cities.
Data are from Table 2-1, *Introduction to Solar Heating and Cooling Design and Sizing*, DOE/CS-0011, 1978.
Table columns are:
  0° Horizontal.
  20° Approximates 4/12 roof slope.
  _° Approximate latitude—usually best for year-round heat collection.
  _° Approximate latitude plus 15°—usually best for winter heating.
  90° Vertical.
States are in alphabetical order.

Phoenix, AZ (33.26° North Latitude)

Month	Avg. Temp. °F	0°	20°	30°	50°	90°
		- -Radiation, Btu/day-ft² - -				
Jan	50.0	1093	1538	1706	1903	1703
Feb	55.4	1502	1943	2092	2223	1809
Mar	59.0	1918	2224	2295	2261	1548
Apr	66.2	2367	2469	2430	2180	1140
May	75.2	2666	2569	2436	2018	827
Jun	84.2	2725	2535	2364	1889	704
Jul	89.6	2400	2275	2143	1759	738
Aug	87.8	2253	2265	2195	1910	934
Sep	82.4	2091	2324	2355	2238	1386
Oct	71.6	1664	2084	2217	2306	1790
Nov	59.0	1248	1736	1916	2121	1864
Dec	51.8	1031	1502	1683	1909	1753

Boulder, CO (40.00° North Latitude)

Month	Avg. Temp. °F	0°	20°	40°	60°	90°
		- -Radiation, Btu/day-ft² - -				
Jan	32.0	740	1104	1358	1471	1355
Feb	33.8	988	1304	1496	1538	1318
Mar	37.4	1478	1758	1869	1797	1375
Apr	48.2	1695	1799	1743	1535	1003
May	57.2	1695	1676	1530	1272	759
Jun	66.2	1935	1860	1652	1332	750
Jul	73.4	1917	1865	1676	1366	780
Aug	71.6	1618	1659	1564	1342	843
Sep	62.6	1518	1716	1753	1625	1168
Oct	53.6	1142	1460	1636	1650	1371
Nov	41.0	818	1185	1434	1535	1392
Dec	35.6	670	1040	1305	1434	1344

Little Rock, AR (34.44° North Latitude)

Month	Avg. Temp. °F	0°	20°	30°	50°	90°
		- -Radiation, Btu/day-ft² - -				
Jan	41.0	729	970	1058	1157	1020
Feb	44.6	964	1184	1254	1306	1051
Mar	51.8	1318	1486	1520	1484	1033
Apr	60.8	1675	1732	1704	1539	872
May	69.8	1944	1886	1801	1529	738
Jun	77.0	2069	1950	1836	1514	681
Jul	80.6	2054	1961	1857	1549	712
Aug	78.8	1900	1913	1858	1633	851
Sep	73.4	1627	1785	1802	1711	1094
Oct	62.6	1274	1553	1639	1689	1314
Nov	51.8	898	1198	1306	1425	1241
Dec	42.8	688	946	1043	1159	1048

Gainesville, FL (29.39° North Latitude)

Month	Avg. Temp. °F	0°	20°	30°	40°	90°
		- -Radiation, Btu/day-ft² - -				
Jan	55.4	1023	1340	1452	1528	1335
Feb	57.2	1351	1647	1737	1785	1386
Mar	62.6	1638	1820	1847	1831	1162
Apr	69.8	1984	2017	1965	1869	888
May	75.2	2157	2052	1936	1782	681
Jun	78.8	2003	1855	1729	1573	590
Jul	80.6	1914	1797	1686	1545	604
Aug	80.6	1870	1846	1774	1666	739
Sep	78.8	1635	1745	1741	1697	970
Oct	71.6	1355	1591	1654	1678	1225
Nov	62.6.	1170	1519	1640	1719	1469
Dec	57.2	935	1250	1364	1445	1298

Davis, CA (38.33° North Latitude)

Month	Avg. Temp. °F	0°	20°	40°	50°	90°
		- -Radiation, Btu/day-ft² - -				
Jan	44.6	581	793	932	968	879
Feb	48.2	942	1207	1359	1386	1163
Mar	51.8	1480	1736	1826	1805	1313
Apr	57.2	1944	2059	1987	1882	1101
May	62.6	2342	2306	2081	1903	921
Jun	69.8	2585	2460	2147	1927	839
Jul	73.4	2540	2455	2172	1965	885
Aug	71.6	2249	2315	2172	2026	1078
Sep	69.8	1833	2083	2130	2074	1384
Oct	62.6	1281	1631	1821	1848	1503
Nov	51.8	795	1107	1313	1367	1236
Dec	44.6	544	772	926	971	905

Griffin, GA (33.15° North Latitude)

Month	Avg. Temp. °F	0°	20°	30°	50°	90°
		- -Radiation, Btu/day-ft² - -				
Jan	42.8	876	1181	1293	1418	1249
Feb	46.4	1112	1372	1456	1517	1214
Mar	51.8	1428	1608	1643	1598	1097
Apr	60.8	1911	1973	1938	1740	950
May	69.8	2124	2050	1951	1640	753
Jun	75.2	2135	2002	1879	1535	668
Jul	77.0	2058	1955	1847	1531	689
Aug	77.0	1925	1929	1869	1632	832
Sep	71.6	1609	1751	1762	1662	1047
Oct	62.6	1369	1664	1753	1800	1384
Nov	51.8	1060	1426	1559	1703	1478
Dec	44.6	773	1061	1170	1297	1165

Table 27. Average total daily solar radiation, continued.

Boise, ID (43.34° North Latitude)

Month	Avg. Temp. °F	Collector tilt angle				
		0°	20°	40°	60°	90°
		- -Radiation, Btu/day-ft²- -				
Jan	30.2	522	784	970	1058	989
Feb	33.8	858	1165	1361	1421	1247
Mar	41.0	1248	1500	1611	1568	1233
Apr	48.2	1789	1936	1908	1708	1148
May	57.2	2161	2167	2001	1674	991
Jun	64.4	2353	2282	2044	1657	923
Jul	73.4	2463	2426	2199	1800	1008
Aug	71.6	2095	2200	2111	1834	1156
Sep	62.6	1679	1964	2057	1947	1440
Oct	51.8	1156	1549	1788	1844	1578
Nov	39.2	666	997	1230	1336	1237
Dec	32.0	452	704	887	980	931

Ames, IA (42.02° North Latitude)

Month	Avg. Temp. °F	Collector tilt angle				
		0°	20°	40°	60°	90°
		- -Radiation, Btu/day-ft²- -				
Jan	19.4	640	972	1208	1320	1230
Feb	24.8	931	1256	1459	1516	1320
Mar	32.0	1204	1424	1513	1459	1134
Apr	48.2	1484	1578	1537	1366	918
May	59.0	1767	1759	1617	1354	818
Jun	68.0	1992	1925	1723	1400	800
Jul	73.4	1973	1933	1749	1438	834
Aug	71.6	1693	1753	1666	1442	918
Sep	62.6	1351	1532	1573	1469	1076
Oct	51.8	1008	1298	1463	1485	1250
Nov	35.6	688	1005	1224	1318	1207
Dec	24.8	526	816	1026	1131	1068

Lemont, IL (41.40° North Latitude)

Month	Avg. Temp. °F	Collector tilt angle				
		0°	20°	40°	60°	90°
		- -Radiation, Btu/day-ft²- -				
Jan	24.8	629	937	1153	1251	1159
Feb	28.4	854	1124	1288	1326	1144
Mar	35.6	1200	1412	1494	1436	1110
Apr	50.0	1436	1520	1476	1308	876
May	59.0	1830	1818	1667	1390	830
Jun	69.8	2036	1964	1752	1418	800
Jul	73.4	1940	1897	1713	1405	812
Aug	71.6	1789	1851	1756	1515	953
Sep	64.4	1414	1602	1644	1532	1116
Oct	53.6	975	1239	1386	1399	1169
Nov	39.2	578	801	950	1007	908
Dec	28.4	482	718	885	964	901

Dodge City, KS (37.46° North Latitude)

Month	Avg. Temp. °F	Collector tilt angle				
		0°	20°	40°	50°	90°
		- -Radiation, Btu/day-ft²- -				
Jan	30.2	953	1417	1737	1829	1706
Feb	33.8	1204	1584	1808	1851	1562
Mar	41.0	1590	1867	1962	1938	1396
Apr	53.6	1988	2099	2019	1908	1100
May	62.6	2073	2033	1832	1677	831
Jun	73.4	2426	2304	2009	1804	798
Jul	78.8	2393	2306	2037	1842	839
Aug	77.0	2146	2196	2052	1910	1014
Sep	68.0	1815	2048	2082	2022	1335
Oct	55.4	1399	1787	1997	2026	1639
Nov	41.0	1031	1487	1792	1875	1709
Dec	32.0	854	1311	1634	1732	1652

Indianapolis, IN (39.44° North Latitude)

Month	Avg. Temp. °F	Collector tilt angle				
		0°	20°	40°	50°	90°
		- -Radiation, Btu/day-ft²- -				
Jan	28.4	541	743	877	913	835
Feb	32.0	788	994	1111	1131	950
Mar	39.2	1148	1324	1380	1361	998
Apr	51.8	1447	1519	1463	1389	848
May	60.8	1808	1784	1624	1497	788
Jun	71.6	2014	1931	1710	1553	759
Jul	75.2	1995	1938	1736	1585	792
Aug	73.4	1789	1836	1728	1619	911
Sep	66.2	1491	1676	1705	1660	1128
Oct	55.4	1078	1357	1507	1528	1247
Nov	41.0	648	885	1039	1079	974
Dec	32.0	478	674	808	846	791

Louisville, KY (38.11° North Latitude)

Month	Avg. Temp. °F	Collector tilt angle				
		0°	20°	40°	50°	90°
		- -Radiation, Btu/day-ft²- -				
Jan	32.0	604	827	974	1013	920
Feb	35.6	851	1070	1191	1210	1007
Mar	42.8	1198	1373	1424	1401	1013
Apr	55.4	1548	1621	1555	1472	880
May	64.4	1898	1866	1688	1550	793
Jun	71.6	2064	1970	1734	1568	747
Jul	75.2	2027	1960	1745	1589	775
Aug	75.2	1835	1875	1755	1639	904
Sep	68.0	1504	1677	1696	1645	1101
Oct	57.2	1117	1391	1533	1549	1248
Nov	44.6	700	947	1106	1146	1024
Dec	33.8	552	782	938	982	914

Table 27. Average total daily solar radiation, continued.

New Orleans, LA (29.59° North Latitude)

Month	Avg. Temp. °F	Collector tilt angle 0°	20°	30°	40°	90°
		- -Radiation, Btu/day-ft²- -				
Jan	51.8	788	992	1061	1106	944
Feb	55.4	954	1117	1162	1182	899
Mar	59.0	1235	1345	1356	1339	857
Apr	68.0	1518	1535	1495	1424	719
May	73.4	1655	1581	1499	1389	599
Jun	78.8	1633	1524	1428	1309	546
Jul	80.6	1537	1451	1369	1263	548
Aug	80.6	1533	1512	1456	1371	647
Sep	77.0	1411	1497	1490	1451	843
Oct	68.0	1316	1543	1604	1626	1189
Nov	59.0	1024	1306	1402	1464	1240
Dec	53.6	729	936	1009	1058	929

Portland, ME (43.39° North Latitude)

Month	Avg. Temp. °F	Collector tilt angle 0°	20°	40°	60°	90°
		- -Radiation, Btu/day-ft²- -				
Jan	23.0	578	889	1114	1223	1149
Feb	24.8	872	1190	1393	1456	1279
Mar	32.0	1321	1600	1727	1684	1326
Apr	41.0	1495	1601	1569	1403	953
May	51.8	1889	1890	1746	1468	889
Jun	60.8	1992	1933	1739	1423	824
Jul	68.0	2065	2033	1848	1526	890
Aug	66.2	1774	1851	1771	1541	989
Sep	59.0	1410	1621	1680	1582	1172
Oct	48.2	1005	1317	1502	1537	1309
Nov	37.4	578	839	1020	1099	1010
Dec	28.4	508	815	1042	1160	1109

Shreveport, LA (32.25° North Latitude)

Month	Avg. Temp. °F	Collector tilt angle 0°	20°	30°	50°	90°
		- -Radiation, Btu/day-ft²- -				
Jan	46.4	832	1096	1191	1293	1122
Feb	50.0	1027	1243	1310	1351	1066
Mar	57.2	1392	1554	1583	1532	1041
Apr	64.4	1719	1762	1726	1544	848
May	71.6	2018	1942	1845	1548	714
Jun	78.8	2003	1875	1758	1437	634
Jul	82.4	2069	1959	1847	1523	674
Aug	82.4	1914	1911	1848	1607	809
Sep	77.0	1528	1649	1654	1552	971
Oct	66.2	1274	1522	1594	1622	1231
Nov	55.4	894	1155	1246	1337	1136
Dec	48.2	729	974	1064	1166	1032

Annapolis, MD (38.59° North Latitude)

Month	Avg. Temp. °F	Collector tilt angle 0°	20°	40°	50°	90°
		- -Radiation, Btu/day-ft²- -				
Jan	33.8	645	904	1078	1126	1035
Feb	35.6	895	1142	1282	1307	1096
Mar	42.8	1253	1448	1510	1489	1083
Apr	53.6	1544	1620	1558	1476	887
May	62.6	1799	1770	1606	1478	770
Jun	71.6	2053	1963	1731	1568	753
Jul	75.2	1998	1935	1727	1574	777
Aug	73.4	1729	1766	1656	1549	867
Sep	68.0	1411	1571	1588	1541	1038
Oct	57.2	1083	1351	1491	1508	1219
Nov	46.4	696	950	1114	1156	1039
Dec	35.6	571	823	997	1048	984

Caribou, ME (46.52° North Latitude)

Month	Avg. Temp. °F	Collector tilt angle 0°	20°	50°	60°	90°
		- -Radiation, Btu/day-ft²- -				
Jan	10.4	504	829	1151	1201	1156
Feb	14.0	846	1219	1540	1570	1417
Mar	24.8	1351	1696	1894	1865	1510
Apr	35.6	1473	1600	1532	1441	1006
May	48.2	1745	1762	1541	1406	887
Jun	59.0	1767	1730	1458	1315	803
Jul	64.4	1874	1861	1591	1441	882
Aug	60.8	1657	1746	1608	1492	991
Sep	53.6	1226	1423	1472	1419	1080
Oct	42.8	773	1016	1196	1199	1038
Nov	32.0	405	584	747	766	711
Dec	17.6	390	655	923	966	942

Amherst, MA (42.15° North Latitude)

Month	Avg. Temp. °F	Collector tilt angle 0°	20°	40°	60°	90°
		- -Radiation, Btu/day-ft²- -				
Jan	24.8	427	588	697	740	674
Feb	26.6	651	825	927	943	808
Mar	33.8	1104	1295	1369	1317	1023
Apr	46.4	1277	1348	1309	1163	790
May	57.2	1587	1578	1452	1221	751
Jun	66.2	1892	1831	1641	1338	774
Jul	69.8	1900	1862	1686	1390	813
Aug	68.0	1620	1675	1592	1379	883
Sep	60.8	1215	1366	1395	1300	955
Oct	51.8	920	1170	1310	1324	1112
Nov	39.2	563	789	941	1002	909
Dec	28.4	456	685	848	927	870

Table 27. Average total daily solar radiation, continued.

East Lansing, MI (42.44° North Latitude)

Month	Avg. Temp. °F	Collector tilt angle 0°	20°	40°	60°	90°
		- -Radiation, Btu/day-ft² - -				
Jan	23.0	423	586	696	741	677
Feb	24.8	736	959	1094	1124	972
Mar	32.0	1082	1270	1343	1292	1006
Apr	46.4	1248	1317	1279	1137	776
May	55.4	1730	1724	1588	1333	811
Jun	66.2	1911	1850	1659	1355	784
Jul	69.8	1881	1845	1674	1382	812
Aug	68.0	1623	1681	1600	1387	891
Sep	60.8	1299	1473	1512	1414	1040
Oct	50.0	891	1131	1266	1279	1076
Nov	37.4	475	642	753	793	713
Dec	28.4	379	544	660	713	662

Great Falls, MT (47.29° North Latitude)

Month	Avg. Temp. °F	Collector tilt angle 0°	20°	50°	60°	90°
		- -Radiation, Btu/day-ft² - -				
Jan	21.2	508	861	1217	1273	1236
Feb	26.6	843	1233	1575	1610	1463
Mar	32.0	1333	1685	1894	1869	1523
Apr	42.8	1579	1732	1672	1576	1104
May	51.8	1929	1957	1719	1570	985
Jun	60.8	2176	2137	1798	1617	960
Jul	68.0	2338	2334	1998	1804	1074
Aug	68.0	1947	2076	1931	1794	1187
Sep	57.2	1487	1775	1878	1820	1395
Oct	48.2	964	1335	1634	1652	1453
Nov	33.8	567	915	1254	1304	1244
Dec	28.4	412	727	1052	1107	1091

St. Cloud, MN (45.34° North Latitude)

Month	Avg. Temp. °F	Collector tilt angle 0°	20°	50°	60°	90°
		- -Radiation, Btu/day-ft² - -				
Jan	10.4	625	1044	1458	1522	1462
Feb	14.0	924	1321	1659	1687	1513
Mar	28.4	1347	1669	1839	1805	1446
Apr	42.8	1557	1688	1607	1508	1040
May	55.4	1837	1850	1606	1461	905
Jun	64.4	1992	1944	1622	1454	860
Jul	69.8	2043	2023	1715	1546	923
Aug	68.0	1808	1903	1743	1612	1053
Sep	59.0	1325	1537	1583	1524	1148
Oct	46.4	887	1172	1383	1386	1194
Nov	32.0	537	807	1058	1089	1017
Dec	17.6	452	756	1060	1109	1075

Lincoln, NE (40.51° North Latitude)

Month	Avg. Temp. °F	Collector tilt angle 0°	20°	40°	60°	90°
		- -Radiation, Btu/day-ft² - -				
Jan	24.8	699	1042	1282	1391	1283
Feb	28.4	939	1239	1420	1461	1256
Mar	37.4	1277	1502	1586	1521	1168
Apr	51.8	1561	1653	1602	1415	935
May	60.8	1826	1809	1653	1373	813
Jun	71.6	2006	1931	1717	1384	777
Jul	77.0	1977	1927	1734	1415	807
Aug	75.2	1870	1931	1826	1567	972
Sep	66.2	1517	1719	1761	1637	1182
Oct	53.6	1196	1549	1750	1775	1483
Nov	39.2	762	1099	1329	1423	1291
Dec	30.2	633	982	1234	1357	1275

Columbia, MO (38.58° North Latitude)

Month	Avg. Temp. °F	Collector tilt angle 0°	20°	40°	50°	90°
		- -Radiation, Btu/day-ft² - -				
Jan	30.2	662	934	1117	1168	1076
Feb	32.0	920	1178	1326	1353	1137
Mar	42.8	1266	1465	1529	1508	1097
Apr	53.6	1594	1675	1611	1527	915
May	64.4	1955	1925	1744	1602	819
Jun	73.4	2102	2009	1770	1602	763
Jul	77.0	2113	2046	1824	1660	806
Aug	75.2	1936	1985	1863	1740	957
Sep	68.0	1649	1859	1892	1841	1236
Oct	57.2	1193	1507	1676	1699	1380
Nov	42.8	817	1150	1370	1429	1299
Dec	32.0	622	914	1118	1178	1114

Reno, NV (39.30° North Latitude)

Month	Avg. Temp. °F	Collector tilt angle 0°	20°	40°	50°	90°
		- -Radiation, Btu/day-ft² - -				
Jan	32.0	862	1308	1620	1713	1623
Feb	35.6	1194	1612	1869	1926	1660
Mar	39.2	1655	1984	2116	2102	1550
Apr	46.4	2182	2340	2276	2162	1263
May	53.6	2447	2421	2192	2008	973
Jun	60.8	2632	2513	2202	1981	871
Jul	68.0	2606	2527	2245	2034	923
Aug	66.2	2381	2468	2328	2176	1160
Sep	59.0	1961	2263	2337	2284	1541
Oct	50.0	1456	1921	2189	2238	1855
Nov	39.2	1021	1530	1880	1981	1847
Dec	32.0	770	1210	1527	1626	1577

Table 27. Average total daily solar radiation, continued.

Trenton, NJ (40.13° North Latitude)

Month	Avg. Temp. °F	Collector tilt angle				
		0°	20°	40°	60°	90°
		- -Radiation, Btu/day-ft² - -				
Jan	32.0	637	923	1119	1203	1102
Feb	32.0	899	1172	1334	1365	1168
Mar	41.0	1264	1480	1559	1491	1141
Apr	51.8	1563	1652	1599	1409	928
May	60.8	1810	1790	1634	1356	800
Jun	69.8	2012	1934	1717	1382	771
Jul	75.2	1990	1938	1740	1417	804
Aug	73.4	1729	1777	1677	1438	898
Sep	66.2	1434	1613	1645	1524	1098
Oct	55.4	1083	1376	1537	1547	1284
Nov	44.6	718	1016	1216	1294	1167
Dec	33.8	571	856	1058	1153	1074

Raleigh, NC (35.47° North Latitude)

Month	Avg. Temp. °F	Collector tilt angle				
		0°	20°	40°	50°	90°
		- -Radiation, Btu/day-ft² - -				
Jan	41.0	876	1229	1461	1523	1378
Feb	42.8	1123	1425	1590	1615	1326
Mar	50.0	1480	1698	1756	1723	1213
Apr	59.0	1741	1811	1722	1620	926
May	68.0	1837	1789	1603	1464	729
Jun	75.2	2099	1984	1726	1550	705
Jul	77.0	1999	1916	1688	1527	720
Aug	77.0	1778	1796	1663	1544	827
Sep	71.6	1421	1554	1547	1547	970
Oct	60.8	1148	1397	1514	1519	1190
Nov	50.0	880	1188	1382	1428	1257
Dec	41.0	743	1059	1272	1331	1227

Albuquerque, NM (35.03° North Latitude)

Month	Avg. Temp. °F	Collector tilt angle				
		0°	20°	40°	50°	90°
		- -Radiation, Btu/day-ft² - -				
Jan	33.8	1134	1667	2027	2127	1950
Feb	39.2	1436	1886	2144	2190	1815
Mar	44.6	1885	2213	2319	2283	1599
Apr	53.6	2319	2439	2325	2183	1182
May	62.6	2533	2461	2181	1972	868
Jun	71.6	2721	2551	2182	1932	754
Jul	77.0	2540	2422	2109	1889	795
Aug	73.4	2342	2376	2194	2028	1011
Sep	68.0	2084	2345	2369	2291	1455
Oct	55.4	1646	2102	2341	2369	1878
Nov	42.8	1244	1789	2146	2240	2011
Dec	33.8	1034	1570	1942	2052	1927

Bismarck, ND (46.47° North Latitude)

Month	Avg. Temp. °F	Collector tilt angle				
		0°	20°	50°	60°	90°
		- -Radiation, Btu/day-ft² - -				
Jan	8.6	581	991	1402	1467	1421
Feb	12.2	924	1353	1728	1765	1599
Mar	26.6	1292	1611	1789	1760	1422
Apr	42.8	1653	1810	1742	1639	1138
May	53.6	2029	2055	1796	1635	1011
Jun	62.6	2161	2117	1773	1591	938
Jul	69.8	2253	2242	1910	1723	1023
Aug	68.0	1907	2023	1870	1734	1139
Sep	57.2	1406	1655	1731	1672	1272
Oct	44.6	1005	1382	1681	1696	1484
Nov	30.2	592	942	1279	1327	1258
Dec	15.8	456	800	1152	1210	1188

Albany, NY (42.40° North Latitude)

Month	Avg. Temp. °F	Collector tilt angle				
		0°	20°	40°	60°	90°
		- -Radiation, Btu/day-ft² - -				
Jan	23.0	456	644	773	828	760
Feb	24.8	673	862	974	995	856
Mar	32.0	1263	1507	1610	1559	1216
Apr	46.4	1340	1420	1381	1229	834
May	57.2	1623	1616	1488	1252	769
Jun	66.2	2220	2147	1919	1553	868
Jul	71.6	1970	1932	1751	1442	840
Aug	69.8	1642	1700	1618	1403	900
Sep	62.6	1148	1287	1312	1222	900
Oct	50.0	946	1211	1363	1382	1164
Nov	39.2	810	1237	1538	1677	1553
Dec	28.4	515	804	1014	1120	1061

Columbus, OH (40.00° North Latitude)

Month	Avg. Temp. °F	Collector tilt angle				
		0°	20°	40°	60°	90°
		- -Radiation, Btu/day-ft² - -				
Jan	30.2	475	639	747	785	705
Feb	32.0	729	915	1019	1029	871
Mar	39.2	1089	1254	1308	1243	949
Apr	50.0	1447	1523	1470	1295	858
May	60.8	1797	1777	1621	1344	794
Jun	69.8	2069	1987	1762	1414	783
Jul	73.4	1995	1942	1743	1418	803
Aug	71.6	1756	1805	1703	1459	908
Sep	66.2	1554	1759	1799	1669	1199
Oct	53.6	1053	1329	1480	1486	1231
Nov	41.0	655	907	1073	1134	1017
Dec	32.0	486	697	844	908	837

Table 27. Average total daily solar radiation, continued.

Oklahoma City, OK (35.24° North Latitude)

Month	Avg. Temp. °F	0°	20°	40°	50°	90°
				Radiation, Btu/day-ft²		
Jan	35.6	939	1329	1587	1656	1501
Feb	39.2	1167	1485	1660	1686	1384
Mar	48.2	1498	1718	1776	1742	1223
Apr	59.0	1833	1909	1814	1706	965
May	68.0	1988	1934	1729	1575	764
Jun	77.0	2319	2187	1891	1690	733
Jul	80.6	2246	2147	1882	1695	762
Aug	80.6	2165	2194	2029	1878	961
Sep	73.4	1782	1978	1985	1916	1232
Oct	60.8	1395	1736	1907	1922	1514
Nov	48.2	1045	1451	1712	1777	1578
Dec	39.2	872	1278	1555	1635	1519

Newport, RI (41.29° North Latitude)

Month	Avg. Temp. °F	0°	20°	40°	60°	90°
				Radiation, Btu/day-ft²		
Jan	28.4	570	828	1007	1085	998
Feb	30.2	850	1116	1278	1314	1132
Mar	35.6	1215	1430	1513	1454	1123
Apr	44.6	1454	1540	1495	1324	886
May	53.6	1800	1788	1639	1367	817
Jun	62.6	1981	1911	1705	1381	784
Jul	69.8	1903	1860	1680	1378	799
Aug	68.0	1653	1705	1615	1394	884
Sep	60.8	1399	1583	1622	1511	1099
Oct	51.8	1005	1280	1434	1449	1212
Nov	41.0	644	912	1094	1166	1057
Dec	32.0	519	785	975	1066	999

Corvallis, OR (44.33° North Latitude)

Month	Avg. Temp. °F	0°	20°	40°	60°	90°
				Radiation, Btu/day-ft²		
Jan	37.4	371	522	627	672	620
Feb	42.8	511	637	710	720	619
Mar	44.6	1034	1227	1309	1271	1003
Apr	50.0	1487	1600	1574	1413	968
May	55.4	1870	1877	1741	1470	900
Jun	60.8	2139	2082	1876	1537	886
Jul	64.4	2467	2439	2220	1827	1036
Aug	64.4	2014	2121	2042	1784	1141
Sep	60.8	1469	1707	1784	1690	1262
Oct	51.8	865	1121	1272	1300	1109
Nov	44.6	515	745	905	976	900
Dec	41.0	298	421	508	547	510

Charleston, SC (32.54° North Latitude)

Month	Avg. Temp. °F	0°	20°	30°	50°	90°
				Radiation, Btu/day-ft²		
Jan	50.0	931	1257	1376	1509	1325
Feb	50.0	1115	1369	1449	1504	1196
Mar	57.2	1443	1619	1652	1603	1093
Apr	64.4	1896	1952	1915	1714	929
May	71.6	2025	1951	1855	1558	721
Jun	77.0	2062	1930	1811	1478	647
Jul	80.6	1925	1827	1726	1432	656
Aug	78.8	1826	1824	1765	1539	787
Sep	75.2	1502	1622	1628	1529	960
Oct	66.2	1263	1511	1583	1613	1229
Nov	57.2	1049	1396	1520	1653	1423
Dec	50.0	795	1086	1194	1321	1181

State College, PA (40.48° North Latitude)

Month	Avg. Temp. °F	0°	20°	40°	60°	90°
				Radiation, Btu/day-ft²		
Jan	28.4	511	709	842	894	811
Feb	28.4	743	942	1057	1072	912
Mar	35.6	1093	1264	1321	1260	966
Apr	48.2	1373	1444	1396	1232	822
May	59.0	1719	1702	1556	1296	775
Jun	68.0	2003	1927	1713	1382	775
Jul	71.6	1944	1895	1705	1392	797
Aug	69.8	1671	1719	1623	1396	877
Sep	62.6	1329	1490	1516	1404	1016
Oct	51.8	1012	1278	1424	1431	1189
Nov	41.0	570	774	906	952	851
Dec	30.2	441	626	753	808	744

Rapid City, SD (44.09° North Latitude)

Month	Avg. Temp. °F	0°	20°	40°	60°	90°
				Radiation, Btu/day-ft²		
Jan	23.0	684	1122	1444	1611	1536
Feb	26.6	1023	1458	1745	1850	1647
Mar	32.0	1469	1817	1985	1954	1551
Apr	44.6	1785	1940	1919	1723	1166
May	53.6	1973	1981	1835	1546	936
Jun	64.4	2180	2120	1908	1559	893
Jul	71.6	2191	2162	1969	1628	946
Aug	69.8	1992	2094	2014	1757	1123
Sep	59.0	1583	1851	1940	1840	1371
Oct	48.2	1156	1568	1822	1889	1627
Nov	35.6	754	1185	1495	1646	1544
Dec	28.4	581	991	1298	1466	1420

Table 27. Average total daily solar radiation, continued.

Oak Ridge, TN (36.01° North Latitude)

Month	Avg. Temp. °F	Collector tilt angle				
		0°	20°	40°	50°	90°
		- -Radiation, Btu/day-ft²- -				
Jan	37.4	611	806	928	957	847
Feb	39.2	876	1079	1184	1196	974
Mar	46.4	1200	1355	1391	1361	963
Apr	57.2	1646	1712	1630	1536	890
May	66.2	1896	1850	1660	1517	754
Jun	73.4	2006	1903	1661	1496	698
Jul	75.2	1918	1842	1629	1478	712
Aug	75.2	1752	1773	1645	1530	827
Sep	69.8	1535	1693	1695	1637	1069
Oct	59.0	1174	1440	1570	1578	1244
Nov	46.4	777	1034	1195	1233	1083
Dec	37.4	592	813	959	998	909

Norfolk, VA (36.54° North Latitude)

Month	Avg. Temp. °F	Collector tilt angle				
		0°	20°	40°	50°	90°
		- -Radiation, Btu/day-ft²- -				
Jan	39.2	766	1069	1269	1322	1200
Feb	41.0	995	1257	1400	1422	1174
Mar	46.4	1371	1574	1631	1602	1141
Apr	57.2	1758	1838	1756	1657	958
May	66.2	1990	1946	1748	1598	790
Jun	73.4	2108	2001	1748	1574	726
Jul	77.0	2027	1950	1725	1564	746
Aug	75.2	1773	1798	1672	1556	846
Sep	71.6	1467	1617	1621	1566	1030
Oct	60.8	1142	1403	1533	1542	1222
Nov	50.0	822	1116	1303	1349	1197
Dec	41.0	678	969	1167	1223	1134

Midland, TX (31.56° North Latitude)

Month	Avg. Temp. °F	Collector tilt angle				
		0°	20°	30°	50°	90°
		- -Radiation, Btu/day-ft²- -				
Jan	42.8	1034	1400	1534	1682	1470
Feb	46.4	1325	1647	1750	1824	1446
Mar	53.6	1756	1991	2037	1980	1330
Apr	62.6	2029	2084	2040	1817	958
May	71.6	2272	2178	2063	1713	743
Jun	78.8	2264	2106	1967	1585	652
Jul	80.6	2261	2131	2002	1635	686
Aug	80.6	2161	2155	2080	1797	865
Sep	73.4	1881	2050	2061	1936	1182
Oct	64.4	1469	1774	1863	1903	1441
Nov	51.8	1185	1588	1732	1885	1618
Dec	44.6	1001	1404	1555	1735	1558

Yakima, WA (46.17° North Latitude)

Month	Avg. Temp. °F	Collector tilt angle				
		0°	20°	50°	60°	90°
		- -Radiation, Btu/day-ft²- -				
Jan	32.0	316	448	568	582	542
Feb	39.2	740	1030	1272	1290	1154
Mar	44.6	1226	1513	1666	1635	1317
Apr	53.6	1719	1884	1813	1705	1179
May	60.8	1900	1920	1675	1525	949
Jun	68.0	2382	2332	1943	1738	1000
Jul	73.4	2121	2106	1793	1618	968
Aug	71.6	2213	2365	2191	2030	1311
Sep	64.4	1428	1680	1754	1694	1285
Oct	51.8	843	1118	1326	1331	1153
Nov	41.0	456	673	874	899	838
Dec	35.6	364	590	816	851	823

Salt Lake City, UT (40.46° North Latitude)

Month	Avg. Temp. °F	Collector tilt angle				
		0°	20°	40°	60°	90°
		- -Radiation, Btu/day-ft²- -				
Jan	28.4	648	948	1157	1248	1147
Feb	32.0	964	1277	1467	1512	1300
Mar	39.2	1347	1592	1688	1621	1245
Apr	48.2	1826	1949	1897	1674	1092
May	57.2	2191	2174	1982	1634	932
Jun	68.0	2540	2438	2151	1705	890
Jul	77.0	2342	2282	2045	1652	902
Aug	75.2	2084	2159	2044	1750	1069
Sep	64.4	1671	1911	1967	1833	1320
Oct	51.8	1233	1604	1816	1843	1542
Nov	39.2	780	1130	1368	1466	1332
Dec	32.0	567	856	1061	1159	1083

Madison, WI (43.08° North Latitude)

Month	Avg. Temp. °F	Collector tilt angle				
		0°	20°	40°	60°	90°
		- -Radiation, Btu/day-ft²- -				
Jan	17.6	564	856	1065	1165	1090
Feb	21.2	812	1088	1260	1309	1143
Mar	32.0	1232	1475	1580	1534	1203
Apr	44.6	1455	1553	1520	1356	921
May	55.4	1745	1742	1608	1353	827
Jun	66.2	2031	1970	1769	1444	830
Jul	69.8	2046	2011	1827	1507	878
Aug	68.0	1740	1811	1729	1503	964
Sep	59.0	1443	1659	1719	1617	1195
Oct	50.0	993	1292	1468	1499	1273
Nov	33.8	555	793	956	1024	937
Dec	23.0	495	783	993	1101	1048

Table 27. Average total daily solar radiation, continued.

Lander, WY (42.48° North Latitude)

Month	Avg. Temp. °F	Collector tilt angle				
		0°	20°	40°	60°	90°
		- -Radiation, Btu/day-ft^2- -				
Jan	19.4	846	1398	1802	2010	1907
Feb	24.8	1182	1684	2013	2128	1880
Mar	32.0	1660	2057	2245	2202	1727
Apr	41.0	2036	2212	2181	1945	1284
May	51.8	2154	2153	1980	1650	968
Jun	60.8	2485	2403	2141	1721	933
Jul	68.0	2386	2342	2116	1727	964
Aug	66.2	2135	2236	2137	1848	1151
Sep	55.4	1708	1988	2074	1955	1434
Oct	44.6	1310	1776	2060	2128	1817
Nov	32.0	872	1360	1707	1871	1740
Dec	21.2	725	1240	1625	1833	1766

GLOSSARY OF SOLAR TERMS

ABSORBER PLATE—Collector part that absorbs incident solar radiation, converts it to heat energy, and transfers the heat to the working fluid or radiates it to the heated space.

ABSORPTANCE—Measure of incident radiation absorbed by a material; fraction between 0 and 1.

ACTIVE SYSTEM—Solar system requiring pumps or fans to move the heat transfer fluid.

ALTITUDE, SOLAR—Angle between the sun's rays and a horizontal surface.

ANGLE OF INCIDENCE—Angle between the sun's rays and a line perpendicular to the surface on which sunlight is falling.

AZIMUTH, SOLAR—Angle between a north-south line and the horizontal projection of the sun's rays.

BACKPLATE—Back of a solar collector; part farthest from sun.

BTU—British thermal unit; unit of energy; amount of energy required to warm one lb of water one degree Fahrenheit.

CFM—Cubic feet per minute; unit of airflow.

COEFFICIENT OF THERMAL EXPANSION—Change in length of a material for each degree change in temperature.

COLLECTOR—Device to receive and absorb solar energy and convert it to heat.

COLLECTOR EFFICIENCY—Ratio of the useful energy gain for a given time period to the solar energy incident on the surface during the same time period.

CONCENTRATING COLLECTOR—Device which focuses direct radiation; obtains energy at higher temperatures than attainable with a flat plate collector.

CONDENSATE—Liquid formed when a vapor condenses.

CONDUCTION—Heat transfer through or between bodies in physical contact—involves no fluid motion.

CONDUCTIVITY—Measure of how readily heat moves through a material.

CONVECTION—Heat transfer by fluid motion.

COVER—Collector part that admits solar radiation to the absorber, shields the absorber from heat losses to the wind, and reduces longwave radiation losses.

DENSITY—Weight of a unit volume of a substance.

DIFFERENTIAL THERMOSTAT—Switch that makes or breaks contact when the temperature difference between two points exceeds or falls below the setpoint.

DIFFUSE RADIATION—Solar energy scattered by particles in the atmosphere; solar energy available on a cloudy day.

DIRECT GAIN—Solar heating by direct exposure to sunlight.

DIRECT RADIATION—Solar energy arriving without diffusion or scattering; also called direct beam radiation.

EMITTANCE—Measure of the tendency of a material to radiate or emit energy of a specified wavelength; fraction between 0 and 1.

ENERGY INTERCEPT AREA—For flat plate collectors with reflectors, the area of an imaginary rectangle perpendicular to the sun's rays, which contains all rays striking the collector and useful reflector area.

EQUINOX—Date when the earth's axis of rotation has a 0# tilt angle toward the sun; day and night are equal length all over the earth; about March 21 and September 21.

EUTECTIC SALTS—See phase change salts.

FIBER BLOOM—Exposure of glass fibers at the surface of FRP—caused by deterioration of the binding resin.

FIXED COLLECTOR—One that does not follow the sun.

FLAT PLATE COLLECTOR—Basic collector in solar heating systems; consists of a dark-colored plate, usually insulated on the bottom and edges, and often covered by one or more transparent covers.

FLUID—Any liquid or gas.

FOSSIL FUELS—Natural fuels formed from prehistoric plants and animals; e.g. coal, petroleum, natural gas.

FPM—Feet per minute; unit for air velocity.

FREESTANDING—Self-supporting; not mounted on or part of another structure.

FRP—Fiberglass reinforced plastic; materials with glass fibers imbedded in a polyester resin.

GLAUBERS SALT—Sodium sulfate decahydrate; latent heat storage material; melting point = 91 F and heat of fusion = 108 Btu/lb.

GLAZING—Transparent solar collector cover; transmits solar radiation and blocks longwave radiation.

GREENHOUSE EFFECT—Trapping heat inside a glass or plastic enclosure, or trapping heat by the earth's atmosphere, by reducing convective and radiative heat loss.

HEADER—Manifold; a larger diameter pipe connecting smaller tubes through a solar collector absorber or other piping system.

HEAT EXCHANGER—Device that transfers heat between fluids without direct fluid contact; usually metal tubes with one fluid inside and the other outside them.

HEAT OF FUSION—Latent heat required to melt a material or the heat released when it freezes; Btu/lb.

HEAT TRANSFER FLUID—Gas or liquid used to move heat within a solar system.

HEATING WATER—Water (or water plus antifreeze) that transfers heat in liquid heating systems.

INCIDENT ANGLE—Angle of incidence.

INFILTRATION—Inward air leakage through cracks and joints, e.g. at windows and doors, caused by wind pressure and differences in indoor and outdoor temperatures.

INSOLATION—Shortwave or solar energy; includes ultraviolet, visible, and infrared radiation; Btu/hr-ft².

LATENT HEAT—Energy absorbed or released by a material when it changes phases (e.g. from solid to liquid); no temperature change is involved.

LATITUDE—Distance north or south of the earth's equator; degrees.

LONGITUDE—Distance east or west of the prime meridian (Greenwich, England); degrees.

LONGWAVE RADIATION—Low energy radiation with wavelengths longer than 3 micrometers; the type of radiation emitted by solar collector absorber plates; thermal radiation.

LOW GRADE HEAT—Energy in materials at low temperatures or only slightly warmer than their surroundings.

NATURAL CONVECTION—Heat transfer caused by the density difference between hot and cold fluids.

NORMAL—Perpendicular; a line at right angles (90°) to a surface.

OPAQUE—Not transparent; does not let light through.

OUTGASSING—Release of gas or vapor from organic materials as they deteriorate.

PASSIVE SYSTEM—Solar system that relies on natural convection of the heat transfer fluid or direct exposure to sunlight.

PHASE CHANGE MATERIAL—Material that stores energy as latent heat.

PHASE CHANGE SALTS—Phase change materials that melt and freeze at temperatures that make them useful for storing heat in solar heating systems.

POTABLE—Fit for drinking.

PRESSURE DROP—Measures fluid friction or resistance to flow and affects required fan or pumping power.

REFLECTANCE—Measure of incident radiation that reflects off the surface of a material; fraction between 0 and 1.

RETROFIT—Adaptation of a building for a technical innovation, such as solar heating.

R-VALUE—Resistance of a material to heat flow; good insulators have high R-values; hr-ft²-F/Btu.

SELECTIVE SURFACE—One for which longwave emittance is much less than shortwave absorptance; a high fraction of incoming solar energy is absorbed, but little heat energy is lost by longwave radiation.

SENSIBLE HEAT—Energy applied to raise the temperature of a material or the energy removed to cool it.

SERPENTINE—Snake-like; having a curving, winding shape.

SERVICE HOT WATER—Potable hot water for washing, cleaning, or cooking.

SHORTWAVE RADIATION—High energy radiation with wavelengths shorter than 3 micrometers; radiation emitted by very hot objects.

SIDEPLATE—The side of a solar collector; supports absorber and cover.

SOLAR CONSTANT—Average amount of solar energy available on a sun-following surface just outside the earth's atmosphere; about 428 Btu/hr-ft².

SOLAR NOON—Time at which the sun reaches its highest point in the sky (zenith) for the day. Solar noon varies from local clock noon depending on season, longitude, and whether Daylight Savings Time is in effect.

SOLAR TIME—Time based on the position of the sun.

SPECIFIC HEAT—Amount of heat required to raise the temperature of a unit mass of material one degree.

STAGNANT—Not running or flowing.

STAGNATION—No fluid movement through a solar collector; high temperatures can develop when the sun is shining.

SUMMER SOLSTICE—Longest day of the year in the northern hemisphere; the first day of summer; about June 21; the northern hemisphere has a maximum tilt toward the sun.

SUN-FOLLOWING—Tracks or follows the sun so the solar angle of incidence is always 0°.

THERMAL BREAK—Insulating material between two heat conductors which reduces conduction heat transfer.

THERMAL ENERGY—Heat energy.

TILT ANGLE—Angle between a horizontal surface and a collector surface.

TRACKING—Able to follow the movement of the sun.

TRANSMITTANCE—Measure of incident radiation that can pass through a material; fraction between 0 and 1.

TROMBE WALL—Concrete, brick, or adobe wall on the south side of passively heated solar structures; solar energy heats the wall during the day and heat is released to the structure at night.

UL—Underwriters' Laboratories.

ULTRAVIOLET—Radiation band with wavelengths from 0.001 to 0.4 micrometers; a high energy component of solar energy; breaks down some rubber and plastic materials.

WINTER SOLSTICE—Shortest day of the year in the northern hemisphere; first day of winter; about December 21; the northern hemisphere has a maximum tilt away from the sun.

SI (METRIC) CONVERSIONS

Multiply to the right: in. x 2.54 = cm
Divide to the left : cm/2.54 = inches

Unit	Times	Equals
Length		
inches	2.540	centimeters (cm)
feet	0.3048	meters (m)
yards	0.9144	meters (m)
miles	1.609	kilometers (km)
Area		
sq. inches	6.451	sq. centimeters (cm^2)
sq. feet	0.09290	sq. meters (m^2)
sq. yards	0.8361	sq. meters (m^2)
acres	0.4047	hectares (ha)
sq. miles	2.590	sq. kilometers (km^2)
Volume		
cu. inches	16.34	cu. centimeters (cm^3)
cu. inches	0.01634	liters (L)
cu. feet	0.02832	cu. meters (m^3)
teaspoons	4.928	milliliters (mL)
fl. ounces	29.57	milliliters (mL)
quarts (liquid)	0.9464	liters (L)
gallons (liquid)	3.785	liters (L)
Mass		
ounce (dry)	0.02835	kilograms (kg)
pounds	0.4536	kilograms (kg)
ton (2000 lb)	907.2	kilograms (kg)
ton (2000 lb)	0.9072	tonne (t)
Velocity		
feet/minute (fpm)	0.005080	meters/second (m/s)
miles/hour (mph)	1.609	kilometers/hour (km/h)
Flowrate		
cu. feet/minute (cfm)	0.0004719	cu. meters/second (m^3/s)
gallons/minute (gpm)	0.00006309	cu. meters/second (m^3/s)

Unit	Times	Equals
Stress/Pressure		
inches of water (60°F)	248.8	pascals (Pa)
inches of mercury (60°F)	3377.	pascals (Pa)
pounds/sq. inch (psi)	6895	pascals (pa)
pounds/sq. foot (psf)	47.88	pascals (Pa)
Heat/Power/Energy		
Btu/hour	0.2931	watts (W)
horsepower	746.0	watts (W)
kilocalorie/second	4184.	watts (W)
Btu/hour · sq. foot	3.154	watts/sq. meter (W/m^2)
Btu/sq. foot	0.01136	megajoule/sq. meter (MJ/m^2)
Btu	0.000293	kilowatt-hour (kWh)
Btu	1055.	joule (J)
Btu·in/h·ft²·deg F	0.1441	watts/meter·kelvin ($W/m·K$)
Btu/pound·deg F	4184.	joule/kilogram·kelvin ($J/kg·K$)
R (deg F·h·ft²/Btu)	0.1761	($K·m^2/W$)
Langley/minute (Ly/min)	696.8	W/m^2
Langley/minute (Ly/min)	221.2	$Btu/h·ft^2$

Temperature
Fahrenheit (F); Celcius (C); kelvin (K)
$$C = (F - 32)/1.8$$
$$F = (1.8 \times C) + 32$$
$$K = C + 273.15 = 273.15 + (F - 32)/1.8$$

Prefixes
To get larger or smaller units, use prefixes.

Prefix	Symbol	Factor	
nano	n	$10^{-9} =$	0.000 000 001
micro	μ	$10^{-6} =$	0.000 001
milli	m	$10^{-3} =$	0.001
centi*	c	$10^{-2} =$	0.01
deci*	d	$10^{-1} =$	0.1
deka*	da		10
hecto*	h	100	$= 10^2$
kilo	k	1 000	$= 10^3$
mega	M	1 000 000	$= 10^6$
giga	G	1 000 000 000	$= 10^9$

*Avoided in technical and scientific writing.
1 ft = 0.3048 m = 304.8 mm = 0.000 304 8 km